STARFINDER

THE COMPLETE BEGINNER'S GUIDE TO THE NIGHT SKY

Carole Stott and Giles Sparrow

LONDON, NEW YORK, MELBOURNE,
MUNICH, AND DELHI

DORLING KINDERSLEY
Senior Editor Peter Frances
Senior Art Editor Maxine Pedliham
Project Editors Ben Hoare, Rob Houston
Project Art Editor Helen McTeer
Editor Miezan van Zyl
Creative Technical Support Adam Brackenbury, John Goldsmid
Production Controllers Rita Sinha, Elizabeth Warman, Erika Pepe
Jacket Design Lee Ellwood, Duncan Turner
Production Editor Phil Sergeant
Picture Research Louise Thomas
Managing Editor Sarah Larter
Managing Art Editor Philip Ormerod, Michelle Baxter

DK DELHI
Editors Suchismita Banerjee,
Kingshuk Ghoshal, Rohan Sinha
Designer Rajnish Kashyap
DTP Designer Pushpak Tyagi, Dheeraj Arora
DTP Coordinator Sunil Sharma
Art Director Shefali Upadhyay

PRODUCED FOR DORLING KINDERSLEY BY
SANDS PUBLISHING SOLUTIONS
Project Editors David & Sylvia Tombesi-Walton
Project Art Editor Simon Murrell

Important notice
Observing the Sun through any kind of optical device can cause blindness. The author and publishers cannot accept any responsibility for readers who ignore this advice.

This edition published in 2010
First published in Great Britain in 2007 by
Dorling Kindersley Limited
80 Strand, London WC2R 0RL
A Penguin Company

2 4 6 8 10 9 7 5 3

UI 002 SD439 March/2010

A CIP catalogue record for this book
is available from the British Library.

ISBN: 978-1-4053-5270-3

Colour reproduction by GRB, London
Printed and bound in China by Hung Hing Offset Printing Company Limited

See our complete catalogue at

www.dk.com

Contents

Star trails
As the Earth rotates, the position of bright stars can be traced in the sky using a long exposure on a camera. In the foreground is the dome of the Canada–France–Hawaii Telescope on the summit of Mauna Kea, Hawaii.-

Whirlpool Galaxy
This Hubble image shows the winding arms of galaxy M51, also called the Whirlpool Galaxy. M51 is visible with binoculars, but its spirals, which are lanes of stars, gas, and dust, can only be seen using professional equipment.

Northern lights
The Aurora Borealis, also known as the northern lights, is the result of solar particles colliding with gases in the Earth's atmosphere. The levels of oxygen and nitrogen in the air determine the colour of this spectacular light show.

Southern lights
This image, taken from the space shuttle *Discovery*, shows the Aurora Australis like a crown of light encircling the South Pole. The Aurora Australis is not as easily viewed as the Aurora Borealis, as it is only visible in some of the remotest parts of the world.

Amateur astrophotography
An amateur astrophotographer, working from his garden in Queensland, Australia, took this image of the Horsehead Nebula. The dark cloud that appears to be shaped like a horse's head is mostly dust.

Meteor shower
Streaks of light created by meteors, also called shooting stars, can be seen during the annual Leonid meteor showers. Leonids are associated with the constellation Leo, from where the meteors appear to originate.

Comet tails
The two tails of comet Hale-Bopp can clearly be seen in this image. The whitish tail is dust, while the blue is the ion tail. Due to its large size, Hale–Bopp was one of the brightest comets of the 20th century.

Lunar eclipse
This composite photo shows multiple exposures taken during a lunar eclipse. A lunar eclipse only occurs during a full Moon and is the result of the Moon passing through the Earth's shadow.

Finding your way

Each fresh pair of eyes looking at the night sky sees a confusion of stars. The myriad pinpoints of light all seem the same and together appear to form a starry sphere around Earth. This imaginary sphere is a key to finding your way about the sky. Soon, you'll discover that brighter stars make patterns, and these act as signposts. They guide us as our view of the Universe changes, and they form a starry backdrop to the planets as they make their stately progress across the sky. Once recognized in this way, the Universe will unfold before your eyes.

New stars
A cluster of newborn stars shines brightly in the centre of one of the most dynamic star-forming regions known. The whole area, known as N66, is in the Small Magellanic Cloud, a companion galaxy to our own Milky Way.

Looking into space

The Universe is a fascinating place brimming over with countless worlds to explore, but the quantity and diversity of the worlds make exploration seem an impossible goal. Nevertheless, there is order within the Universe. Objects are grouped into types, and a grasp of the scale of the Universe adds structure and form. With this basic understanding, it is possible to begin an enthralling journey of discovery.

Eye in the sky
The planets, stars, and galaxies have been revealed to us by telescopes and space probes. The Hubble Space Telescope has been a constant eye on the Universe since 1990.

To the edge of the visible Universe
The Universe is here separated into five steps, moving farther away from our home planet. From Earth, we move to the Moon, then the Solar System, the Milky Way, and the Local Group of galaxies. Beyond this, however deep we look into space we find more galaxies.

WINDOW ON THE UNIVERSE

The sky of our home planet, Earth, is our window on the Universe. As we look away out into space, we see some of its objects. In the daytime, there is just one large object, the Sun. It is our local star, and its brilliance fills the sky.

On a cloud-free night, the dark sky is full of starry pinpoints of light. With the naked eye we can see more than 2,000 of these objects, and with binoculars the number rises to over 40,000. Looking carefully, we can discern that some are not bright points of light but disc-shaped planets, while the fuzzy patches we see can be galaxies full of stars.

The objects in the Universe are so remote that it is difficult to compare their sizes and distances. They all appear to be an equal distance from Earth. In reality, though, they are not only at varying distances from us, they are also vastly different distances from each other.

Our “neighbourhood” includes the Sun, Moon, and planets. Beyond is the realm of the stars, the Milky Way Galaxy. And outside of this are more galaxies overflowing with stars.

Earth: our home planet
A blue ball of rock 12,756km (7,926 miles) across, Earth is the only place in the Universe known to have life; and uniquely, much of it is covered by liquid water. As Earth spins and moves in space, we are able to look out to different parts of the Universe.

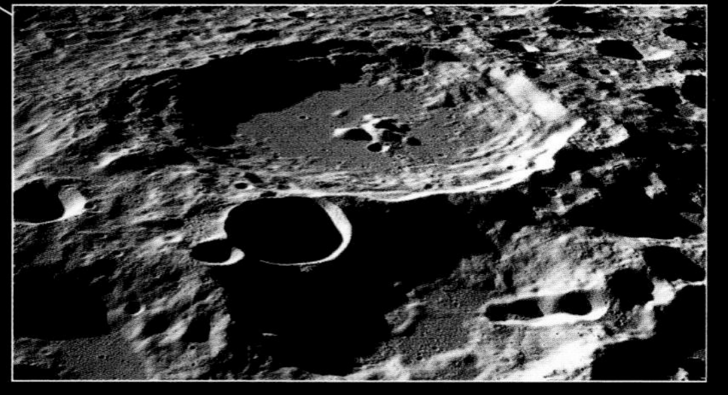

The Moon
Closest to Earth, and a familiar feature of its sky, is the Moon. At its brightest, it is the next brightest object in our sky after the Sun. This ball of rock accompanies Earth in space, travelling around us every 27.3 days. It is a dry, dead world covered in impact craters formed billions of years ago. The Moon is the only place that man has walked apart from Earth.

The Solar System
Earth and the Moon are part of the Solar System (above top), which is a volume of space that includes billions of objects. The central, most massive object is the Sun, and all other members of the system travel around it. Next biggest are the eight planets, and the giant of these is Jupiter (above). Comets and asteroids are more numerous but smaller

MEASURING DISTANCE

The Universe is so vast that the units of measurement used on Earth, such as the kilometre, quickly become inadequate. They are used to describe the size of the planets within the Solar System, as well as distances between them, but beyond the Solar System, the unit commonly used is the light year (ly). One light year is 9.46 trillion (million million) km (5.88 trillion miles) and is the distance that light travels in a year. Light moves at 299,792km per second (186,282 miles per second), and nothing in the Universe travels faster. The Andromeda Galaxy is said to be 2.5 million light years away, because it takes that length of time for its light to reach us; this means we are seeing the galaxy as it was 2.5 million years ago. By contrast, the Sun's light takes just eight and a half minutes to reach Earth. In the chart below, the first division represents 10,000km (6,200 miles). Each further division marks a 10x increase in scale on the previous one.

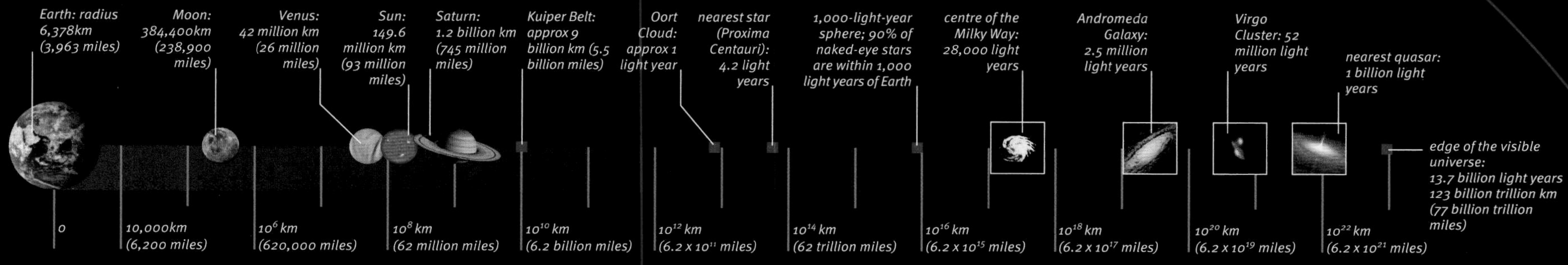

he Milky Way Galaxy
e Sun is just one of billions of stars in the Milky Way alaxy (above top). The stars are in a disc shape, with a ncentration of stars in the centre, and "arms" of stars irralling out from it. All the stars we see in the night sky e in the Milky Way, along with clusters of stars (above) d giant clouds of gas and dust where new stars form.

The Local Group of galaxies
The Milky Way is one of a group of more than 40 galaxies that exist together in space. They are collectively called the Local Group (above top). The Milky Way and the Andromeda Galaxy (above) dominate the group. The Andromeda is about 2.5 times as wide as the Milky Way and is the most distant object normally visible to the naked eye.

Universe of galaxies
In whatever direction we look and however deep we peer into space, there are galaxies (above top) – an estimated 100–125 billion of them. They exist in clusters (above) that are strung together into superclusters, which are the biggest structures in the Universe. Prominent among these are the Hercules and Centaurus superclusters.

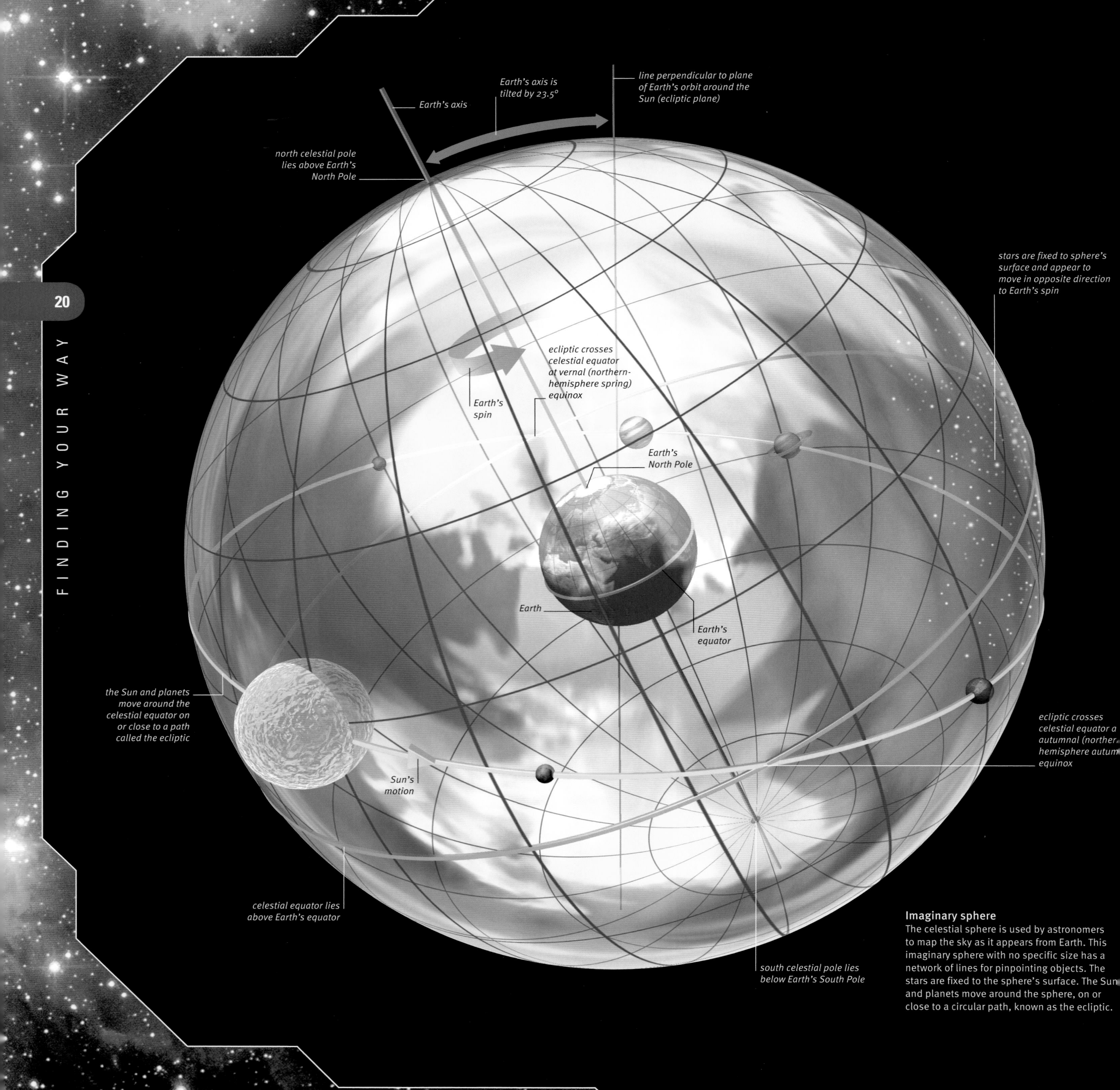

Imaginary sphere
The celestial sphere is used by astronomers to map the sky as it appears from Earth. This imaginary sphere with no specific size has a network of lines for pinpointing objects. The stars are fixed to the sphere's surface. The Sun and planets move around the sphere, on or close to a circular path, known as the ecliptic.

The starry sphere

Looking out from Earth we see a Universe full of stars. The stars are so far away that they all appear to be the same distance from us. As they move across the sky, they do so together, keeping their relative positions. The stars appear to be fixed to the inside of a giant globe enveloping Earth. Although it is a distortion of reality, this imaginary globe, called the celestial sphere, is a useful astronomical tool.

THE CELESTIAL SPHERE

The celestial sphere is important for anyone who wants to become familiar with the night sky. It is used by all observers, from beginners to experienced astronomers. It helps in understanding how location and time and date of observation determine what is in the sky. The sphere's network of lines pinpoints the positions of stars, and helps when navigating around the sky. The celestial equator, concentric with Earth's equator, divides the sphere into the northern- and southern-hemisphere sky. North and south celestial poles are above Earth's poles.

The distant stars are fixed on the sphere, and as Earth spins they appear to revolve around the poles. The much closer Sun and planets move against the backdrop of fixed stars. The Sun's path, the ecliptic, crosses the celestial equator at two points, known as the equinoxes. The Moon and planets follow paths close to the ecliptic.

YOUR VIEW OF THE CELESTIAL SPHERE

An observer's latitude determines which portion of the celestial sphere they see. Most people see all of one celestial hemisphere and part of the other, but not all of this is visible at once. As Earth makes its daily spin and yearly orbit around the Sun, new regions of the sphere come into view.

KEY TO SPHERES
- Horizon
- Observer
- *Stars always visible*
- *Stars visible at some time*
- *Stars never visible*

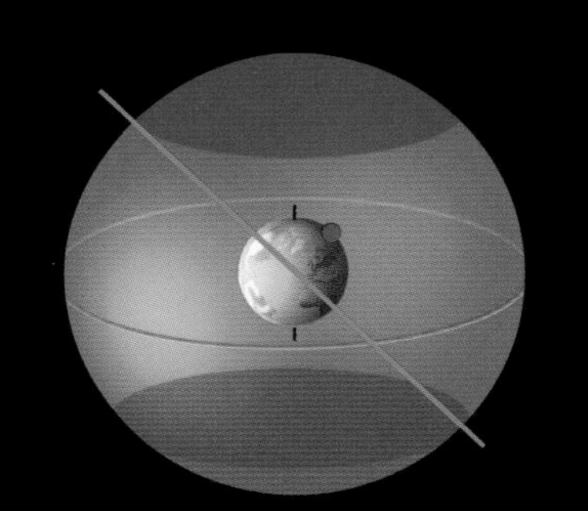

View from the North Pole
To an observer based at Earth's North Pole, only the stars in the northern half of the celestial sphere are visible.

View from mid-latitudes
Stars close to the observer's celestial pole are always visible. Those further from the pole are seen as Earth spins and orbits.

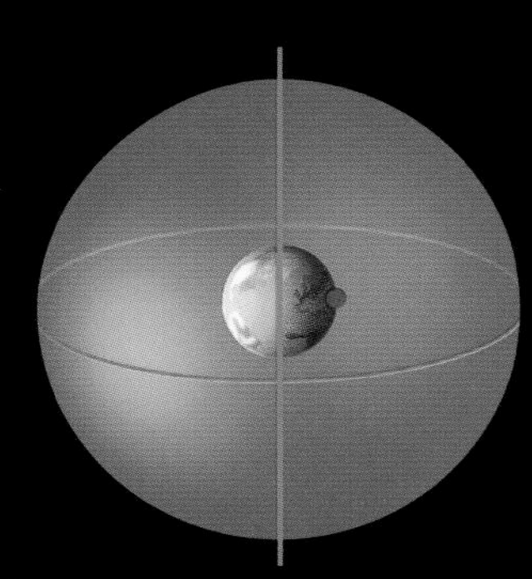

View from the equator
All parts of the celestial sphere are visible. The celestial equator is overhead, and the celestial poles are on opposite horizons.

CHANGING VIEW

From any one location on Earth it is possible to see only part of the celestial sphere at a particular time. This view changes during the evening as the Earth spins. Earth turns from west to east, and so the stars move across the sky from east to west. The view also changes over the course of a year (see pp.22–23). For instance, stars that are in the winter daytime sky can be seen in the evening sky later in the year when the Sun has moved along the ecliptic and its starry backdrop has changed.

An observer's location determines not only which part of the celestial sphere is visible, but also how the stars move across the sky. Except at the equator, observers will always have part of the sky around the celestial pole above the horizon. Stars circle around the pole; the farther you are from the pole, the more circumpolar stars you will see. The position of the celestial pole in your sky corresponds to your latitude on Earth. For instance, for observers at 40°N, the north celestial pole is located 40° above their northern horizon.

ORION FROM USA

ORION FROM JAPAN

The same sky
Observers at the same latitude but on opposite sides of the world (on different longitudes) see the same sky. The constellation of Orion is clearly visible to observers in Arizona, USA, at 11pm (left). Several hours later, once Earth has turned, the same view of Orion is seen at 11pm in Tokyo, Japan (right).

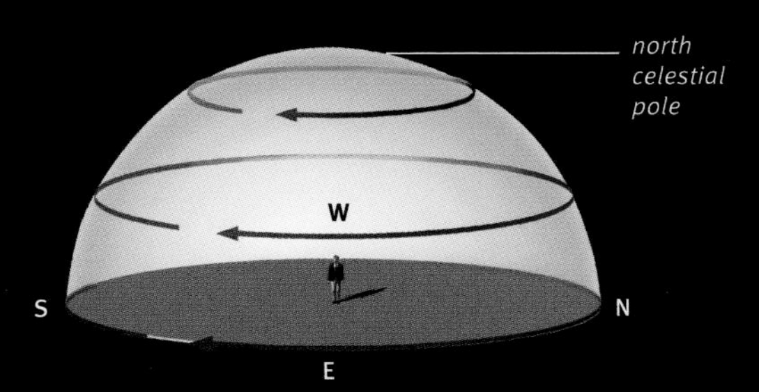

Motion at the poles
At the North Pole, the stars circle overhead, around the north celestial pole. They circle anticlockwise. At the South Pole, the stars circle in the opposite direction, clockwise.

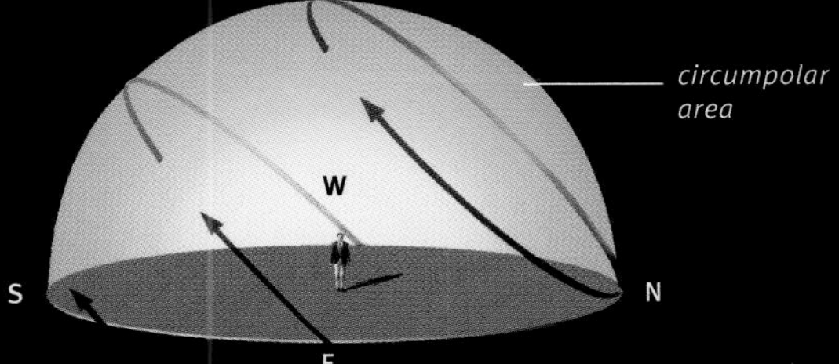

Motion at mid-latitudes
At mid-latitudes, the stars rise above the eastern horizon, cross the sky obliquely, and set below the opposite, western horizon. Also, some circumpolar stars are always in the sky.

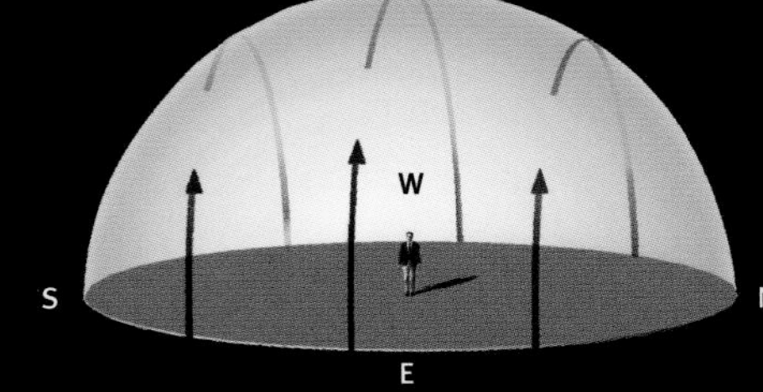

Motion at the equator
From the equator, stars are seen to rise vertically from the eastern horizon, then move across the sky over the observer's head, before setting in the west.

The changing sky

Our view into the Universe is constantly changing. As Earth makes its daily spin, the Sun leaves the sky, darkness falls, and the distant stars come into view. Also, through the course of a year, the starry sky seen from a fixed point on Earth will change. The Sun, Moon, and planets follow their paths against the background of these stars. Particular star patterns – the zodiac constellations – help us locate them and track their progress.

YOUR CHANGING VIEW

One side of Earth is always facing the Sun, which lights up the daytime sky, while the other is facing the night-time stars. Earth spins around on its axis, and as it makes its daily spin, an observer at a fixed location is facing towards both the Sun and the stars in any 24-hour period.

WHY THE VIEW ALTERS

If Earth did not move along its path around the Sun, it would alternately face the Sun and the same piece of starry sky. But because Earth travels around the Sun, it faces progressively different stars at the same time on successive nights during the course of a year. Although this change is hardly noticeable from night to night, as the weeks stretch into months, the difference in the sky becomes more apparent. Observers in the northern hemisphere looking south and those in the southern hemisphere looking north will each see new constellations (see pp.24–25) with the changing seasons. By contrast, the northern sky for the northern-hemisphere observers and the southern sky for the southern-hemisphere observers undergo less change. These views contain the appropriate celestial pole, and the circumpolar stars that move around it are always in view.

As Earth makes its orbit, the starry backdrop to the Sun also changes. Like the Moon and planets, the Sun is seen against the zodiac band of sky (see facing page). Northern-hemisphere observers will see this stretching across the sky to their south, and southern observers see it in the sky to their north.

1 APRIL, 8PM

8 APRIL, 8PM

15 APRIL, 8PM

Night-time sky change
With the exception of those at the North and South Poles, observers see different stars in their sky during the course of a year. An observer at 50°N sees the constellation of Orion lower and lower in the sky as April progresses. By the end of the month, it has almost disappeared below the horizon.

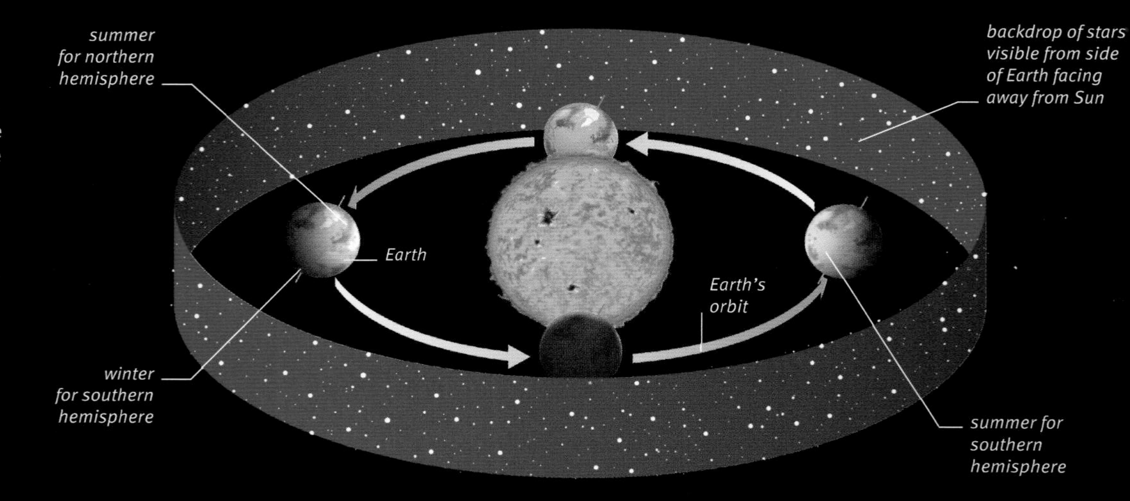

Seasonal sky change
One side of Earth faces the Sun, while the other side faces the stars. If Earth was fixed in space, it would face the Sun and the same piece of starry sky alternately as it makes its daily spin. But Earth is not fixed: it orbits the Sun. As a result, the stars we see in the summer months are different from those we see in winter.

6 AUGUST

7 AUGUST

11 AUGUST

Daytime sky change
As Earth makes its yearly orbit around the Sun, the position at which the Sun rises above the eastern horizon and sets below the western horizon changes. The three images above record sunset at a location 50°N on different days in August. The Sun sets farther and farther towards the south as the summer progresses.

DIVIDING THE SKY

At first glance, stars seem like indistinguishable pinpoints of light randomly scattered across the sky. Quite quickly, however, it is noticeable that some stars are brighter than others, and when these are linked together, recognizable patterns emerge. Astronomers have been using such patterns, the constellations (see pp.24-25), for about 4,000 years to navigate around the night sky. Today, the sky surrounding Earth is divided into 88 internationally recognized constellations, just over half of which depict mythological people and creatures. The constellation Orion depicts the eponymous Greek hunter; the constellation Scorpius is the scorpion that killed him. They range in size and complexity. Hydra, the water snake is the largest, and Crux, the southern cross, the smallest.

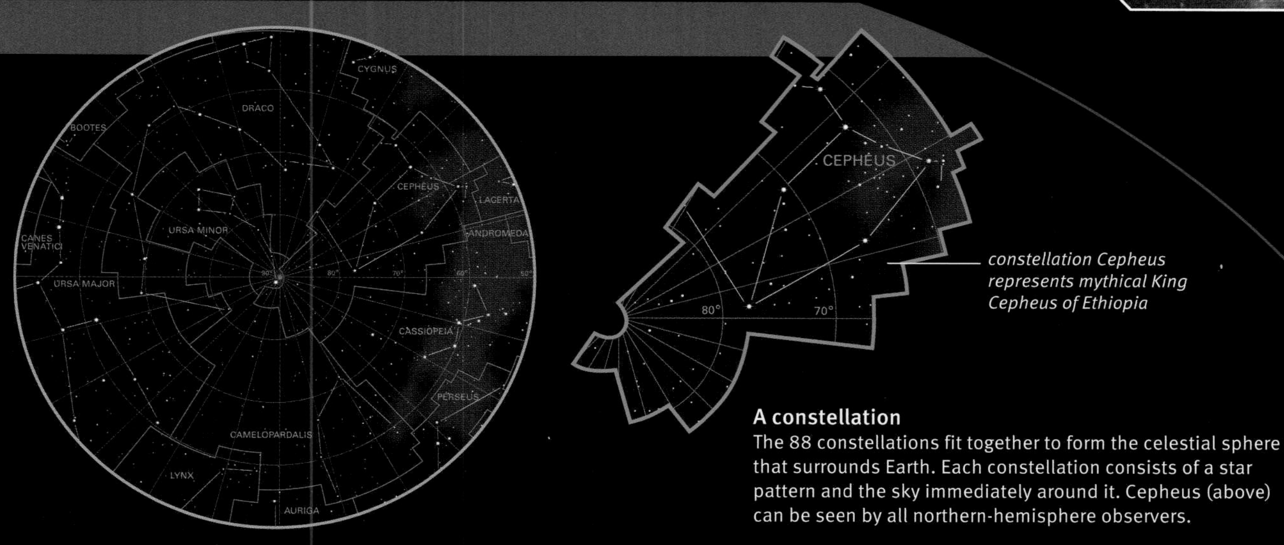

A constellation
The 88 constellations fit together to form the celestial sphere that surrounds Earth. Each constellation consists of a star pattern and the sky immediately around it. Cepheus (above) can be seen by all northern-hemisphere observers.

STARRY BACKDROP

Some of the first stars to be grouped into recognizable patterns were those that form the background to the Sun as it moves across the sky. The Sun's path against the backdrop of stars is called the ecliptic. And the band of sky centred on the ecliptic is known as the zodiac.

CIRCLE OF ANIMALS

The name zodiac comes from the Greek for animal, and it refers to the patterns made by the stars within this band of sky. With the exception of Libra, which is a set of scales, the traditional zodiac is a group of creatures. The zodiac constellations are, in order around the sky: Aries, Taurus, Gemini, Cancer, Leo, Virgo, Libra, Scorpius, Sagittarius, Capricornus, Aquarius, and Pisces.

The Sun completes one circuit of its ecliptic each year. It takes approximately one month to move through each of the zodiac constellations. However, this is not the full picture. Traditionally, the zodiac is described as consisting of 12 constellations. There is, though, a 13th, Ophiuchus, and the Sun spends more time crossing this constellation than it does traversing its neighbour Scorpius.

The zodiac band of sky also forms the starry backdrop to the planets and the Moon. Their paths take them close to the ecliptic – sometimes to its north, and sometimes to its south.

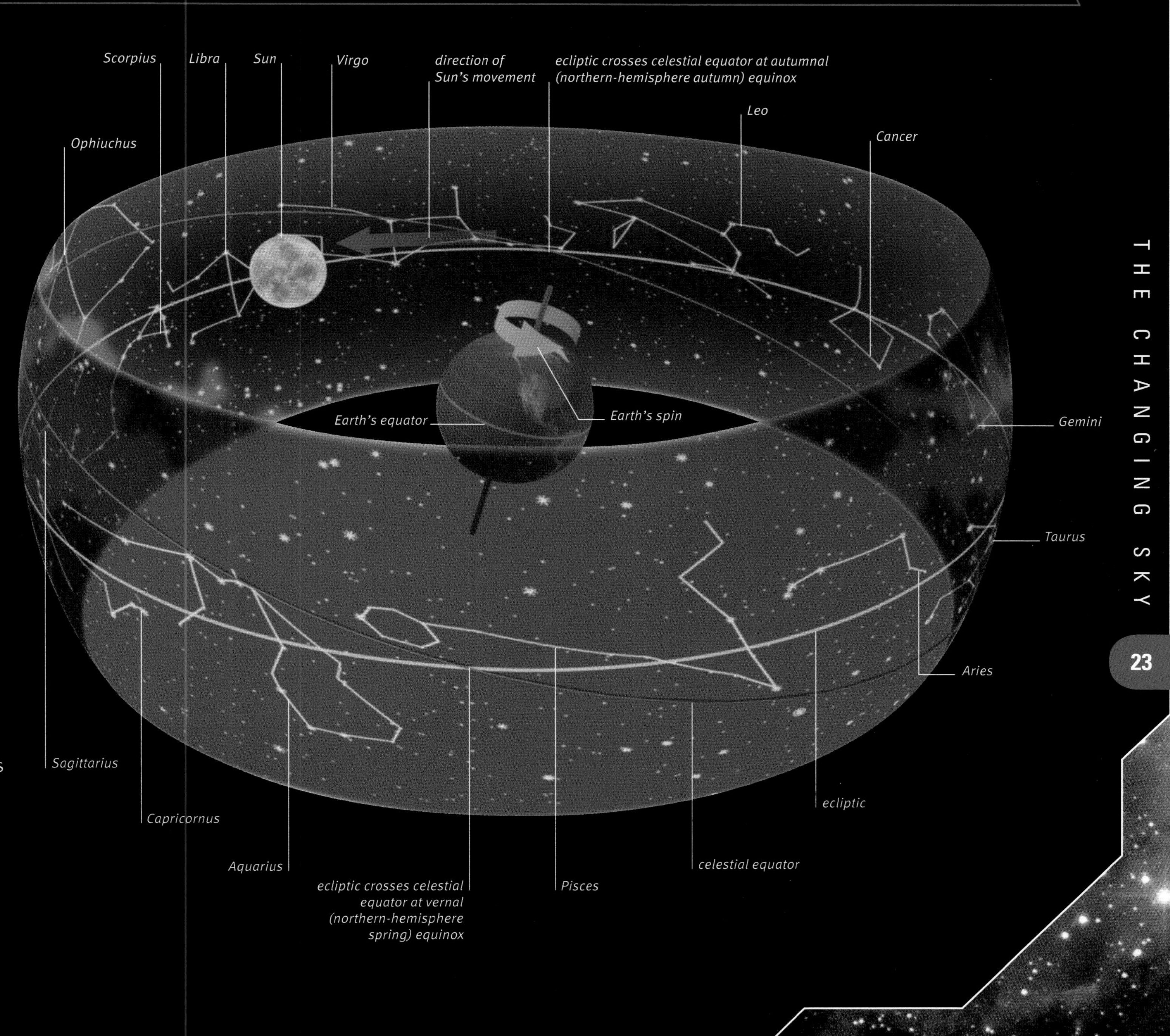

The zodiac
The band of the celestial sphere that forms the backdrop to the movement of the Sun, Moon, and planets is called the zodiac. It is centred on the ecliptic, the path that the Sun traces out month by month as the Earth orbits around it. There are 13 constellations in the zodiac: the 12 traditional zodiacal constellations and Ophiuchus.

A piece of sky

The celestial sphere is divided into areas called constellations, each of which contains a figure produced by joining the brightest stars in dot-to-dot fashion. These figures are purely products of the human imagination – the stars have no real relationship to one another in space. The stars, as well as other objects within a constellation, are identified by name or number according to an agreed system of rules.

Sirius
The brightest night-time star of all is Sirius (below centre), in the constellation Canis Major. Its magnitude is –1.44. At 8.6 light years away, it is also one of the closest stars to us. It gives out about 20 times more light than the Sun.

BRIGHTNESS

It is easy to observe that the stars in the sky vary in brightness. The first star-watchers soon realized this, and so they classed the stars according to their brightness level. Their system was later formalized to produce what is known as the apparent-magnitude scale used today.

Apparent magnitude is an indication of how bright a star appears when viewed from Earth. This is not the same as the star's real brightness, its luminosity. Each star is given a number, known as its magnitude to indicate its brightness: the smaller the number, the brighter the star. The brightest stars have negative magnitude values.

The apparent-magnitude scale is also used for other objects – the full Moon is magnitude –12.5, for example. The brightness of a planet varies depending on its distance from the Sun and Earth; at its brightest, Venus measures –4.7.

TEN BRIGHTEST NIGHT-SKY STARS

The stars listed below are in order of brightness. Only the four brightest have negative values, and as the magnitude number increases, the star becomes fainter. All stars of magnitude 6 and brighter are visible with the naked eye, so each of these stars is easily seen.

1. **Sirius**
   Canis Major, –1.44
2. **Canopus**
   Carina, –0.62
3. **Rigil Kentaurus**
   Centaurus, –0.28
4. **Arcturus**
   Boötes, –0.05
5. **Vega**
   Lyra, 0.03
6. **Capella**
   Auriga, 0.08
7. **Rigel**
   Orion, 0.18
8. **Procyon**
   Canis Minor, 0.40
9. **Achernar**
   Eridanus, 0.45
10. **Betelgeuse**
    Orion, 0.45

RELATIVE DISTANCE

The stars on the celestial sphere all appear to be at the same distance from Earth, but the reality is quite different. The stars are scattered through space with vast expanses between them. Each star is typically 7 light years away from its nearest neighbour. And they are all so far from us that they appear to be pinpoints of light rather than huge globes.

The distance of the naked-eye stars we see, for instance, varies hugely. As we look at them, we see stars that are just a few light years from us and others that are thousands of light years away. This means that the star patterns of the constellations are illusions. They are a two-dimensional view of stars within a three-dimensional volume of space. The stars are not related at all, but far apart. Very often they are farther from each other than they are from Earth.

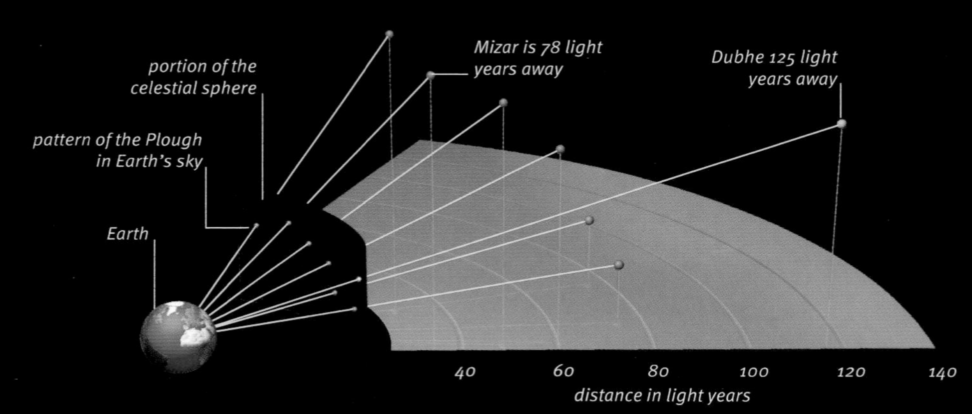

Line-of-sight effect
The stars in the constellation Ursa Major trace out the shape of a bear in Earth's sky. The stars that form the tail and rump of the bear are known as the Plough, or Big Dipper. The stars are at vastly different distances from Earth, and seen from elsewhere in space, they would make a totally different pattern.

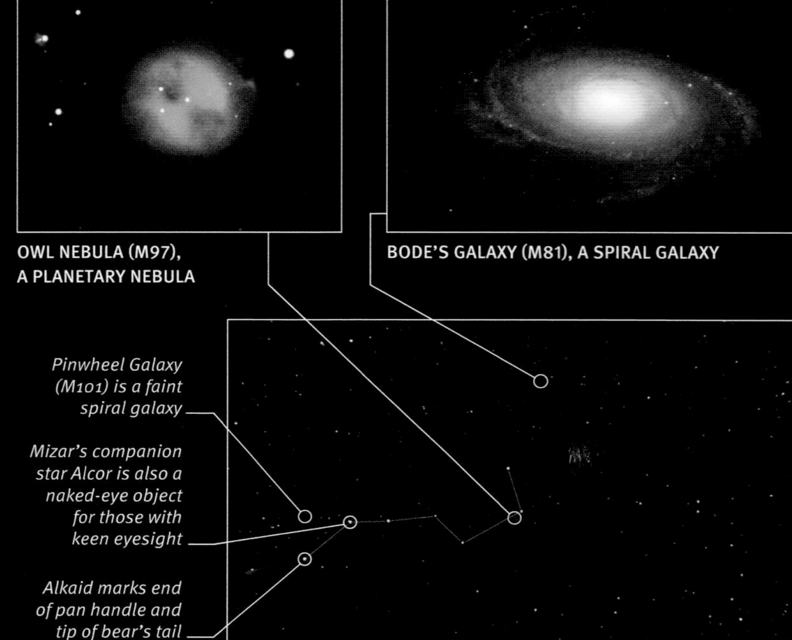

The Plough
Also known as the Big Dipper, the Plough (left of centre) consists of seven bright stars. Three stars form the handle, and four more the basin of a saucepan in profile. Once found, the shape can be used as a guide to other objects. The two brightest stars are known as the pointers, because they point the way to the Pole Star, Polaris.

PATTERNS IN THE SKY

Astronomers worldwide use not only the same set of 88 constellations, but also the same ways of locating and identifying objects within them.

Naming and numbering

Individual stars may be identified by names, numbers, or letters. The brightest stars in a constellation have a Greek letter: alpha (α) is usually the brightest; the next brightest, beta (β); and so on. Once the Greek alphabet is used up, small Roman letters (a,b,c, etc.) are allocated.

Many of the bright stars have names, a large number of which are Arabic in origin. The vast majority, however, are unnamed. Fainter stars, although often within naked-eye visibility, have a number allocated according to their position within the constellation. Stellar objects such as star clusters, nebulae, and galaxies have a catalogue number from one of the many volumes that list such objects. NGC numbers, for example, come from the New General Catalogue.

Asterisms

Some stars make a distinctive pattern in the sky but are not a constellation; they are a separate pattern within a constellation, or are made up of stars from more than one constellation. Such a pattern is known as an asterism. The Plough, or Big Dipper, possibly the best-known pattern in the northern-hemisphere sky, is a prime example. Four stars that form the lower torso of Hercules are known as the Keystone; and the head, neck, and shoulders of Leo is known as the Sickle.

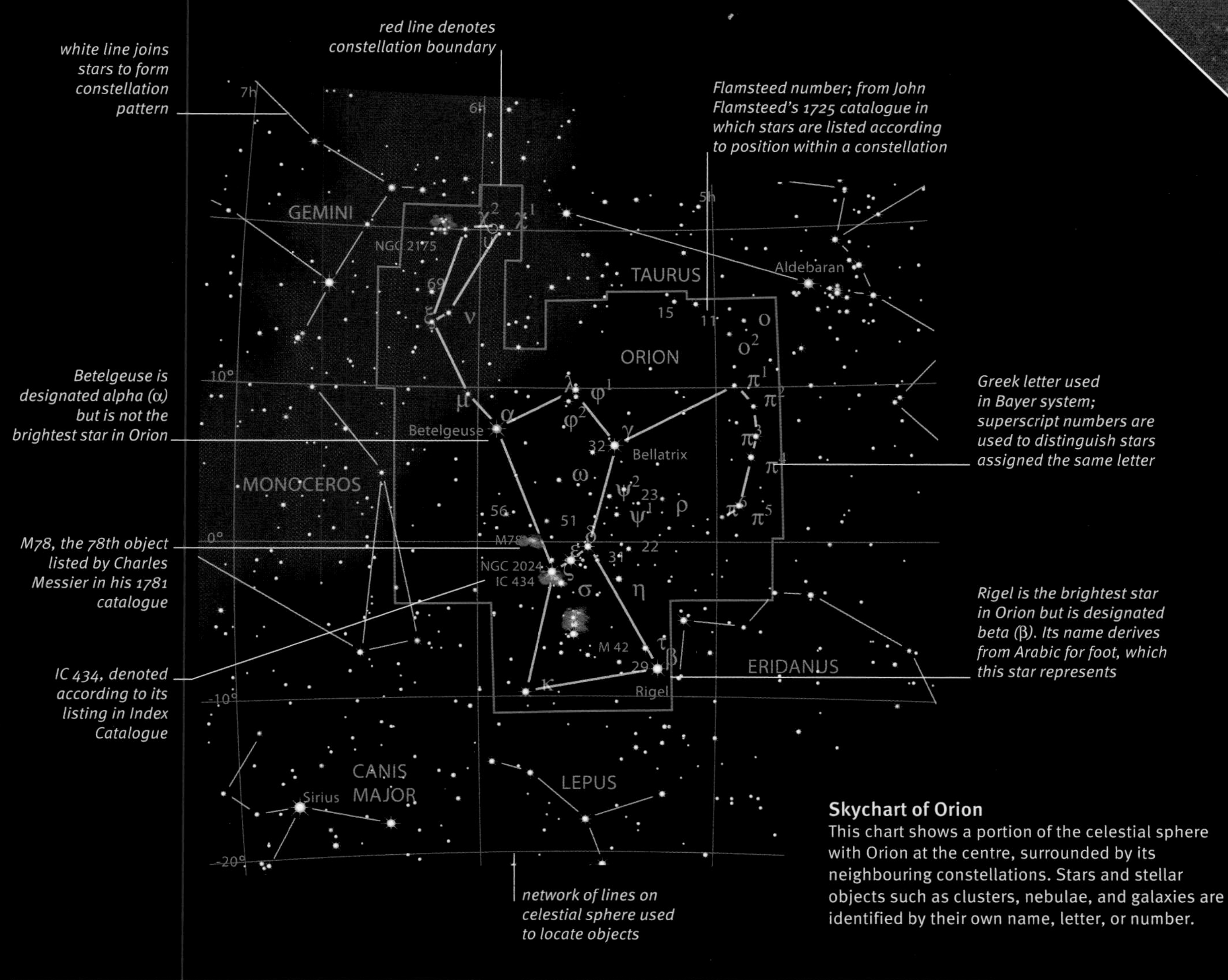

Skychart of Orion
This chart shows a portion of the celestial sphere with Orion at the centre, surrounded by its neighbouring constellations. Stars and stellar objects such as clusters, nebulae, and galaxies are identified by their own name, letter, or number.

Orion in the night sky
One of the most easily seen constellations in the night sky, Orion is an excellent target for novice star-watchers. Its simple, distinctive shape traces the figure of the mythical hunter. It is seen in both the northern- and southern-hemisphere skies

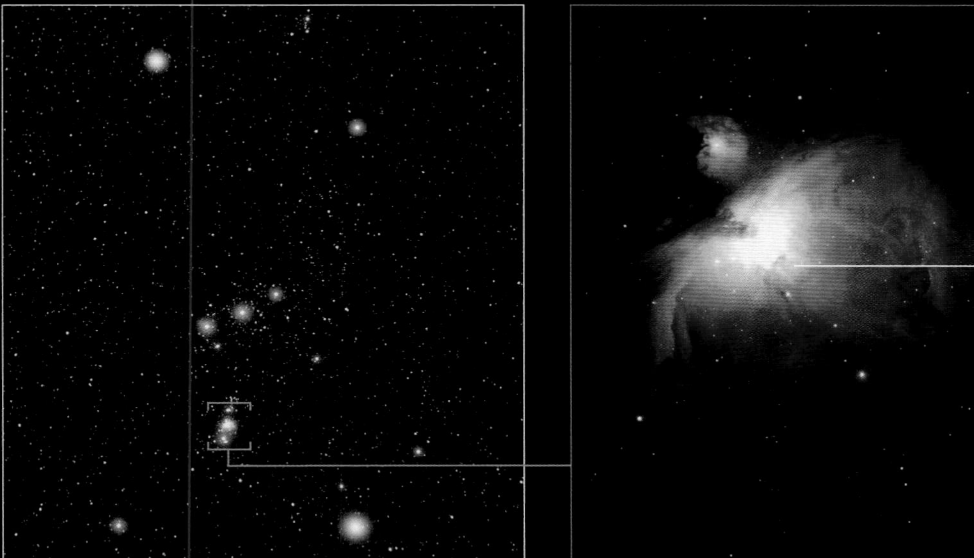

Orion's belt and sword
One of Orion's shoulders is marked by the red supergiant Betelgeuse (top left), while blue supergiant Rigel (lower right) is the constellation's brightest star. Three stars form the belt, from which hangs a sword of stars and glowing gas and dust

Orion Nebula
The Orion Nebula, or M42, is the brightest nebula in the night sky. It is an enormous star-forming cloud of gas and dust some 30 light years across and 1,500 light years away. It is illuminated by newly born stars deep within

Your view of the Solar System

The nearest celestial objects to Earth are members of the Solar System – a group of eight planets and smaller objects that orbit the Sun. The planets and our Moon emit no light of their own but shine by reflecting sunlight. From Earth we can see five of these planets with relative ease. As the planets, Moon, and Earth travel around the Sun, our view changes, at times creating spectacular sights for us on Earth.

VIEWING THE PLANETS

Of the eight planets, Earth is the third in distance from the Sun. Mercury and Venus are closer and, for this reason, are called the inferior planets. The five other planets – Mars, Jupiter, Saturn, Uranus, and Neptune – are all farther away than Earth and are called the superior planets. The five visible to the naked eye are Mercury, Venus, Mars, Jupiter, and Saturn, and their existence has been known of since humans first looked into space. Uranus's brightness suggests that it is just within the range of naked-eye visibility, but it is difficult to see. Uranus was only discovered in 1781, and then through a telescope. Neptune, farthest from the Sun, is too faint to see with the naked eye.

All eight planets follow paths around the Sun, and one complete circuit of the Sun is called an orbit. As each planet orbits, and time passes, the positions of the planets relative to Earth change. The view that we have from Earth of a particular planet depends on these relative positions. For instance, when a planet is on the opposite side of the Sun to Earth, it is lost from view. The planet is still present in the daytime sky, but it is invisible because it is drowned out by the glare of the Sun. For this reason, Mercury and Venus are also invisible when situated between the Sun and Earth. Specific juxtapositions of the Earth, the Sun, and the planets have names. These act as a quick indicator of whether or not a planet is visible. They can also signify the phase, brightness, size, and time of visibility of a planet.

The close orbits of Mercury and Venus to the Sun mean that these two planets are never far from the Sun in the sky. They are best seen when at locations that are known as greatest elongation. At this point, they are at their maximum angle to the east or west of the Sun. The superior planets are best seen at opposition, when they are on the opposite side of the Earth from the Sun. In this position, a planet is close to Earth and appears particularly large and bright. At opposition, a planet is visible all night long. It is found to the south at midnight by observers in the northern hemisphere, and to the north at midnight by observers in the southern hemisphere.

RETROGRADE MOTION

The planets make their stately progress against the starry backdrop from west to east. However, planets are occasionally seen to move in the opposite direction, from east to west, with their paths seeming to loop or zigzag across the sky. This backwards movement, known as retrograde motion, is nothing more than an illusion created by our viewpoint on Earth. It is achieved when the faster-moving planet Earth overtakes a slower-moving superior planet, such as Mars. As Earth moves along its orbit, the superior planet is seen from a different perspective. Superior planets go into retrograde motion around the time of opposition.

MARS SEEN OVER SEVERAL MONTHS

Moon and Venus
The planets and the Moon all move within the zodiac band of constellations and so can often be viewed together. Here, the Moon and Venus are seen in the early evening sky.

Planetary positions
As Earth and the seven other planets orbit the Sun, their relative positions change. The best time to see Mercury and Venus is when they are at maximum elongation; Mars, Jupiter, and Saturn are best seen at opposition. None of the planets is visible at inferior and superior conjunction.

superior conjunction of superior planet
superior conjunction of inferior planet; planet is in full phase and is not visible from Earth
greatest western elongation: planet appears as crescent in morning sky
greatest eastern elongation: planet appears as crescent in evening sky
inferior planet's orbit
inferior conjunction; inferior planet is between Earth and Sun in its new phase and is not visible from Earth
Earth
superior planet's orbit
opposition of superior planet

Mars looping the loop
The retrograde, or backwards, movement of a planet across Earth's sky is an illusion caused by our viewpoint from Earth. It is achieved when the faster-moving planet Earth overtakes a superior planet – in this case, Mars, as each moves from west to east (right to left in this diagram). As Earth overtakes on the inside, Mars seems to perform a zigzag across the sky.

path of Mars across sky
Mars's orbit inclined to ecliptic plane
ecliptic plane
Earth
Mars
Sun
Earth's orbit

ALIGNMENTS

The planets and the Moon are always seen within the same band of sky. So it is not surprising that two or more are often seen close together at the same time. Such an alignment is called a conjunction. Should one object obstruct the view of another, more distant object, it is known as an occultation. The Moon regularly occults background stars. When the Moon or a planet passes in front of a star it is temporarily lost from view. In the case of the Moon, which has no atmosphere, the disappearance and later reappearance of the star happens instantly. A grazing occultation is when the star skims the upper or lower limb of the Moon. In this case, the occulted body is lost from view only as it disappears behind lunar mountain peaks along the limb. An occultation of a bright planet by the Moon occurs 10 or 11 times a year. Occultations of one planet by another – for example, an occultation of Jupiter by Venus – occur only a few times a century.

OTHER CELESTIAL PHENOMENA

When the Moon obstructs our view of the Sun it is also, technically, an occultation, but this phenomenon is better known as an eclipse. During a solar eclipse, the Moon stops the Sun's light reaching Earth. When the Moon moves into Earth's shadow, we witness a lunar eclipse.

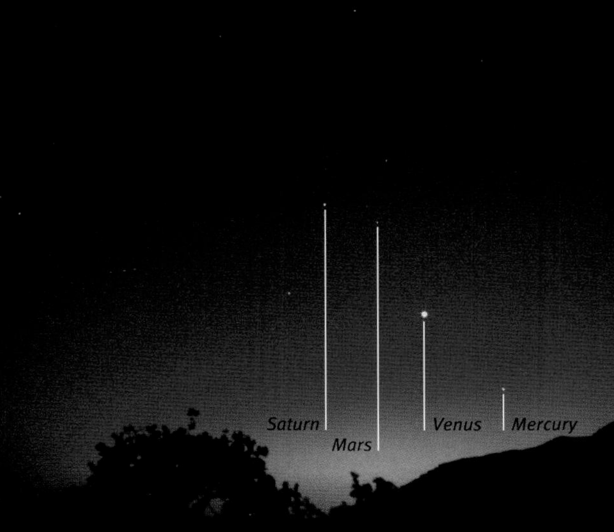

Planetary alignment
In April 2002, soon after the Sun had set, a line of five planets could be seen with the naked eye. Nearest to the horizon was Mercury (bottom right), above and to its left was brighter Venus, and up and to the left again were Mars and Saturn. Bright Jupiter (top left) completed the view.

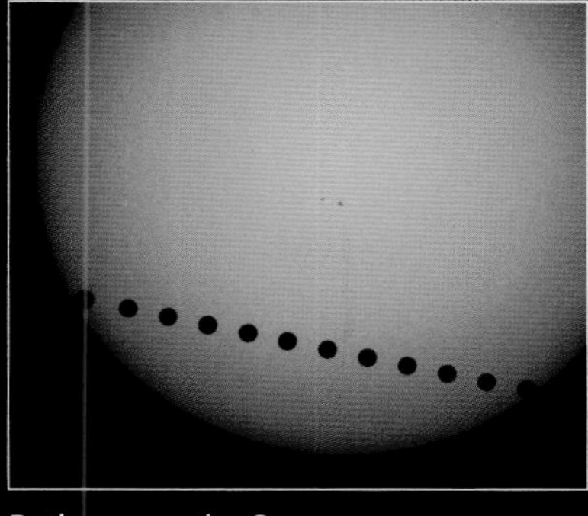

Path across the Sun
This composite image of the 2004 transit of Venus charts the progress of the planet across the Sun's disc. It took more than six hours for the planet to cross the Sun. Transits of Venus occur in pairs, and the partner to this one is due in 2012. The next pair is due in 2117 and 2125.

When a smaller body crosses the disc of a larger one, it is known as a transit. The inferior planets Mercury and Venus cross the Sun's disc when directly between Earth and the Sun. But transits do not occur every time Mercury and Venus pass between Earth and the Sun, because the planets usually pass above or below the Sun's disc. Transits of Venus occur in pairs, separated by more than 100 years, while transits of Mercury are more frequent, occurring about 12 times each century.

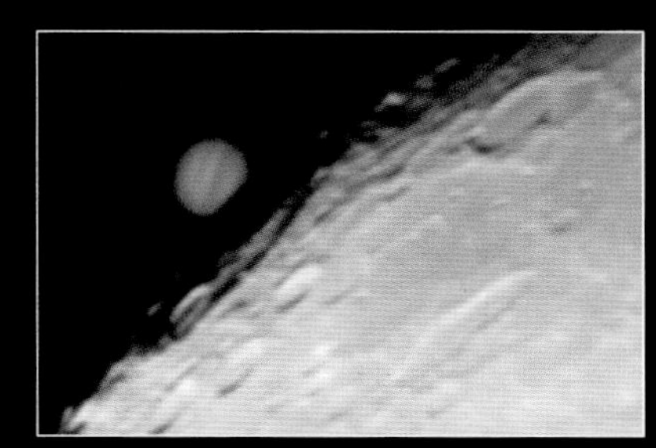

Occultation of Jupiter
This picture shows the January 2002 occultation of Jupiter, which is about to disappear from view behind the Moon. The two objects are along the same line of sight from Earth. Such an alignment often lasts long enough for an occultation to occur each lunar month until the two move out of alignment.

MECHANICS OF ECLIPSES

The Moon is 400 times smaller than the Sun and 400 times closer to Earth, so when the three bodies are directly aligned, the Moon's disc blocks the Sun from view. This is a solar eclipse. The shadow cast by the Moon falls on Earth, and within the shadow, day turns to night as darkness falls. A total solar eclipse typically lasts for 3–4 minutes. In a lunar eclipse, when the Moon is in Earth's shadow, totality can last up to 1 hour 47 mins. There are on average four eclipses each year – two solar and two lunar.

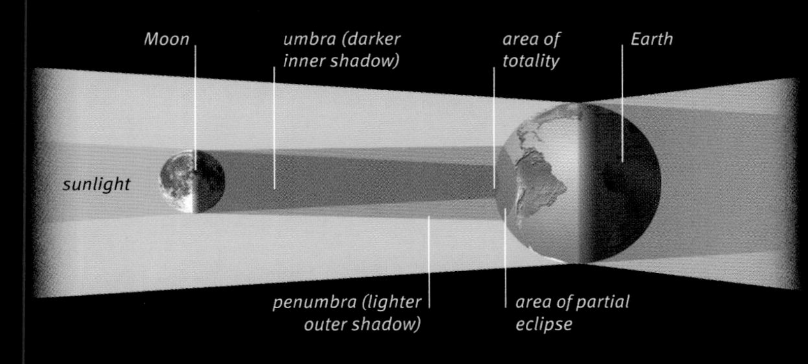

Eclipse of the Sun
When the Moon is directly between the Sun and Earth, it blots out the Sun and casts a shadow on Earth. Anyone within the umbra, the darker inner shadow, sees a total eclipse. Outside of this is a wider area of lighter shadow, the penumbra, from within which a partial eclipse is seen. An annular eclipse occurs when the Moon is farther from Earth than average, and its disc is too small to cover the Sun's disc totally.

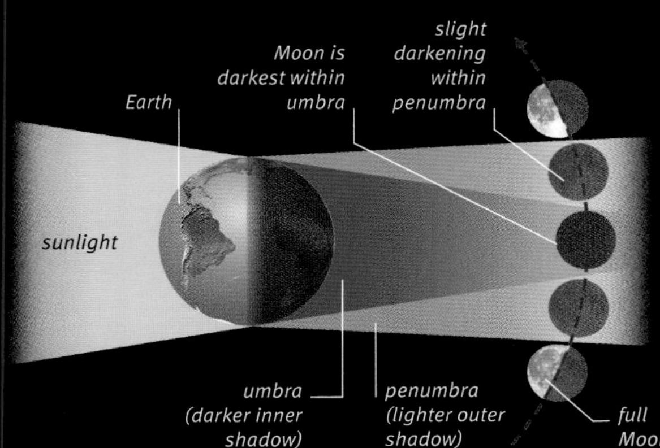

Eclipse of the Moon
The Sun, Earth, and Moon are aligned, and Earth stops sunlight reaching the Moon. When the whole of the Moon is in the umbra, it receives no light and is totally eclipsed. When only a portion of the Moon passes through the umbra, a partial eclipse can be seen.

The Milky Way and beyond

The stars that surround Earth all belong to the Milky Way Galaxy, which is an enormous disc-shaped collection of stars, as well as gas and dust. The Sun and Earth are in a spiral arm of stars about two-thirds out from the centre of the disc. It is an ideal position to observe the galaxy's stellar objects – its stars, clusters, and nebulae. We can also look beyond the Milky Way into a Universe full of galaxies.

Fornax Cluster
Galaxies are not randomly spread though space; they exist in clusters huge distances apart that form superclusters. The Milky Way belongs to the Local Group, which, along with clusters such as the Fornax Cluster, forms the Local Supercluster.

LOOKING AT THE MILKY WAY

The Milky Way is disc-shaped with a bulging centre. Within the disc are spiral arms of stars. Stars also exist between the arms, but because the arm stars are young and bright, these are the ones that shine out. The centre is packed with young and old stars. In total, there are about 500 billion stars.

From our position on Earth we can look into the disc and along its plane, either towards the centre or away from it and out of the galaxy. The many stars in the disc appear as a river of milky light, hence the name Milky Way. The view to the centre is seen in Sagittarius. A view through the plane and out of the galaxy is seen, for example, in the path above Orion's head. We see the other stars in our night sky (those not in the milky path) by looking perpendicular to the disc plane, up or down into the disc.

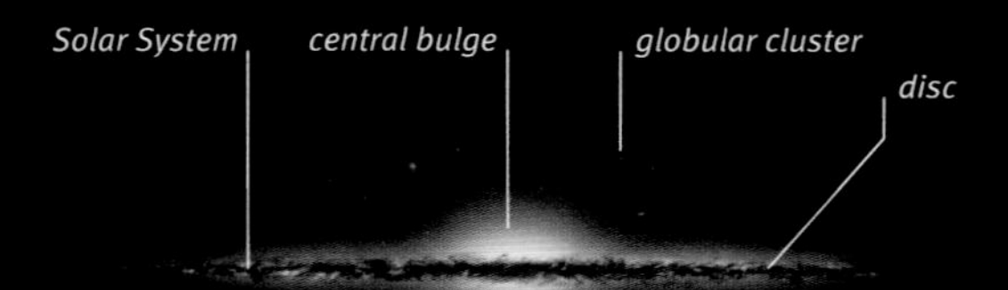

Edge-on view
The Milky Way measures 100,000 light years from side to side, and its disc is roughly 4,000 light years thick. Young stars within the arms give the disc a blue-white tinge, while the bulge has a yellow tinge because it contains older stars. The disc is surrounded by a sparsely populated spherical halo of individual old stars and about 200 globular clusters.

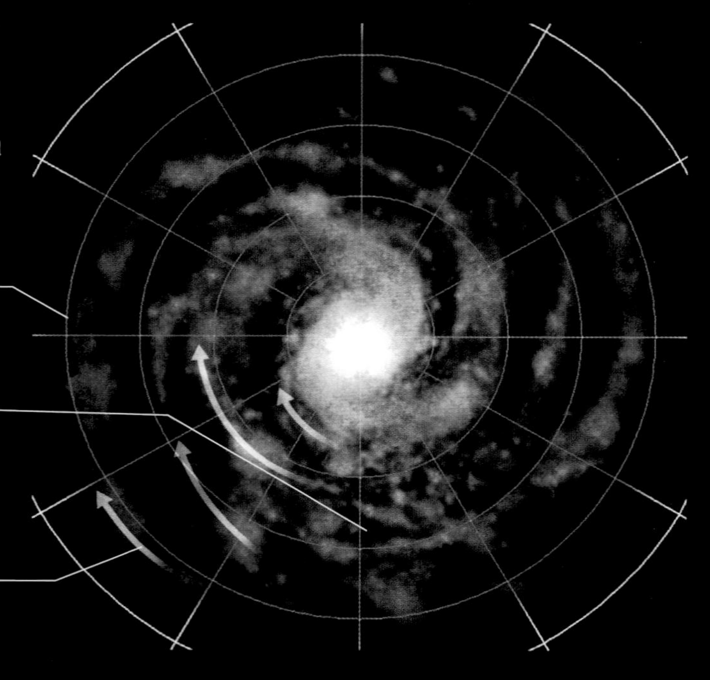

Bird's-eye view
This artist's impression shows what we think the Milky Way looks like from outside and above the disc. Gravity keeps the stellar objects in the Milky Way together. They do not travel as a solid disc but follow individual paths around the galaxy's centre. They move at about the same speed; the closer to the centre, the shorter the orbit time. The Sun's orbit lasts 220 million years.

Milky path of light
When we look into the plane of the Milky Way's disc, we see its stars as a path of light. This path is broadest and brightest within the constellation of Sagittarius, but the view is dappled with dark c of dust that hide more distant stars. One dust patch is known as the Dark Horse (above right of c The horse's legs point to the right edge of the image, and its head towards the top edge.

OBJECTS IN THE MILKY WAY

Most of the galaxy's visible material consists of stars at various stages in their life cycles: young, newly formed ones; middle-aged ones like the Sun; and older red giants and planetary nebulae. The rest is vast clouds of interstellar gas and dust.

STARS, CLUSTERS, AND NEBULAE

A star is a huge ball of hot, glowing gas that is held together by gravity. Like humans, each star is unique, differing from each other in size, brightness, colour, age, and mass – that is, the amount of gas making the star.

A star's size, temperature, and colour change with time. An individual star is described by the stage it has reached in its life cycle. Betelgeuse, for instance, is a red supergiant; and a planetary nebula is an old star that has pushed off its outer layer of gas.

Some stars seen from Earth are double stars. Such stars may not be related, but they look close because of our line of sight from Earth. Other doubles, such as Algol, exist together in space; stars such as this are known as binaries. Stars whose brightness varies are variable stars.

Stars form in clusters from gas and dust, and often they will drift apart. Open clusters consist of young, newly switched-on stars; globular clusters consist of older stars. A cloud of interstellar gas and dust is called a nebula. Nebulae either shine brightly or appear as a dark patch against a brighter background.

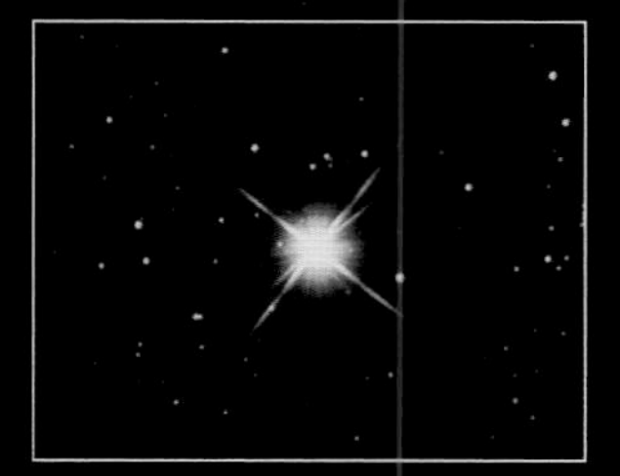

BETELGEUSE (RED SUPERGIANT)

ALGOL (BINARY)

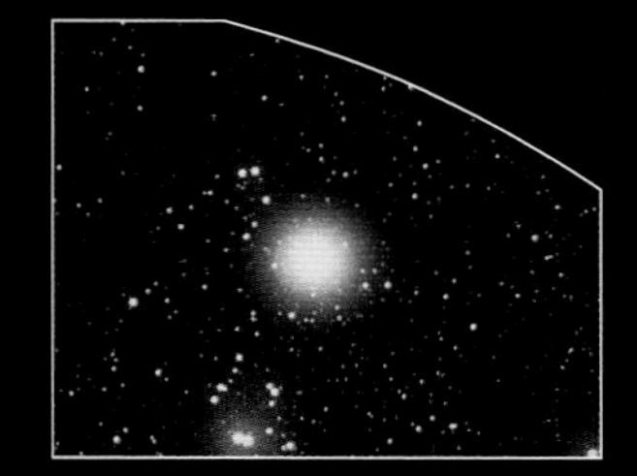

ALDEBARAN (RED GIANT, VARIABLE)

BUTTERFLY CLUSTER (OPEN CLUSTER)

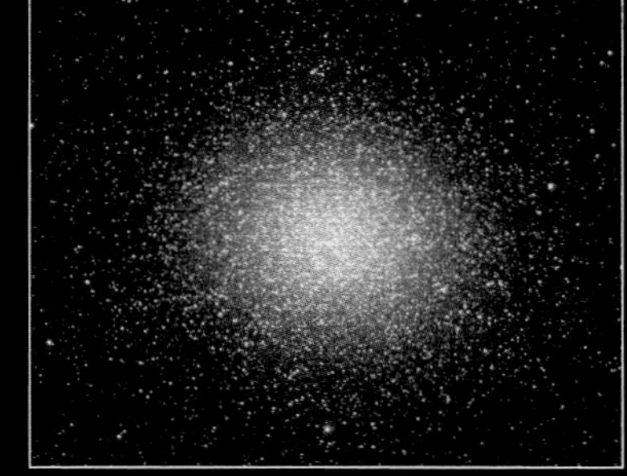

OMEGA CENTAURI (GLOBULAR CLUSTER)

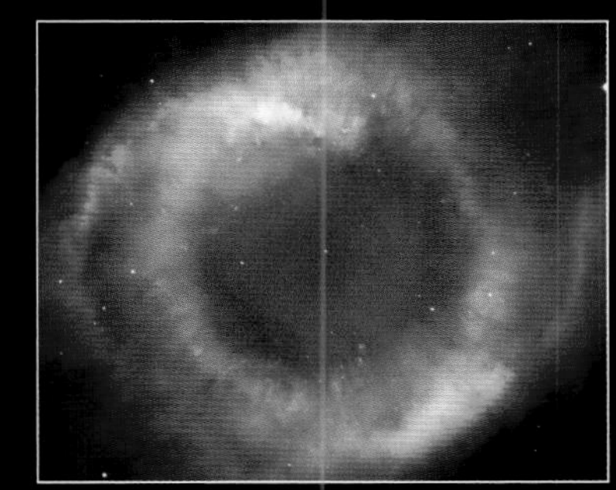

HELIX NEBULA (PLANETARY NEBULA)

LAGOON NEBULA (BRIGHT NEBULA)

HORSEHEAD NEBULA (DARK NEBULA)

BEYOND THE MILKY WAY

When we look out from Earth and beyond the stars of the Milky Way, we see galaxies. They exist in every direction we look. It is estimated that there are 100–125 billion of them. Each is a vast group of stars held together by its own gravity. A single one contains billions or trillions of stars, as well as interstellar gas and dust. Galaxies vary in width from a few thousand light years to more than a million light years. They are so remote that even the nearest look like faint blurs of light to the naked eye.

Galaxies are classified according to shape. Spirals are disc-shaped with a central bulge and spiralling arms. A spiral galaxy with a bar-shaped centre and arms winding out from those bar ends is known as a barred spiral. Ellipticals are ball-shaped, along the lines of a football, a rugby ball, a flattened ball, or something in between. Irregular galaxies have no obvious defined shape or form.

Andromeda Galaxy
The closest major galaxy to the Milky Way, Andromeda is the largest member of the Local Group. To the naked eye, this vast spiral galaxy looks like an elongated smudge of light.

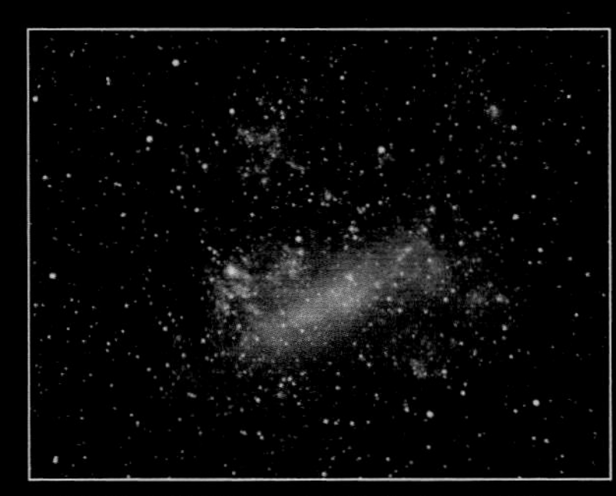

LARGE MAGELLANIC CLOUD (IRREGULAR)

NGC 1300 (BARRED SPIRAL)

M87 (ELLIPTICAL)

Getting started

Anyone can observe the night sky. Simply wait until it is dark, then go outside and look up at the sky. It is thrilling to see the distant stars, but even more so if you can identify them. Simple preparations, such as choosing where to observe from and knowing what you might see, make all the difference. And the successful identification of a few stars each evening will soon grow into a broad knowledge of the sky.

Getting your bearings
A compass will help you get your bearings if you are away from familiar surroundings. Use it in conjunction with the planisphere to locate objects in the night sky.

CHECKLIST

- Take warm clothing, even on summer nights.
- Wear a hat in winter; fingerless gloves are useful.
- Carry binoculars around your neck for ease of use.
- Keep your torch handy: its soft light keeps your eyes adapted to the dark.
- Use a reclining chair, if possible, for extra comfort.
- Take food and drink if observing for a long time.
- Make sure your horizon is clear of obstructions.
- Set up your planisphere before going into the dark.

PLANNING AHEAD

The choice of an observation site dictates the quality of the sky. Stars, planets, nebulae, and galaxies are best seen in a dark sky away from houses and street lights. If the sky is clear and moonless, about 300 stars are visible from cities. Only the brightest shine out, which to the absolute beginner can be an advantage: it is the brightest stars that make the constellation patterns. A darker, village sky will yield about 1,000 stars; and the darkest country location, about 3,000. Here the constellation patterns are not so clear, and dim objects are easier to see.

Once you have chosen a location, think about what you need; the checklist (left) will help. Collect everything together before going out. A lucky few may not even need to leave the house. At the right location, bright stars, the Moon, and planets can all be observed from a window. Simply turn off the lights and look out and up.

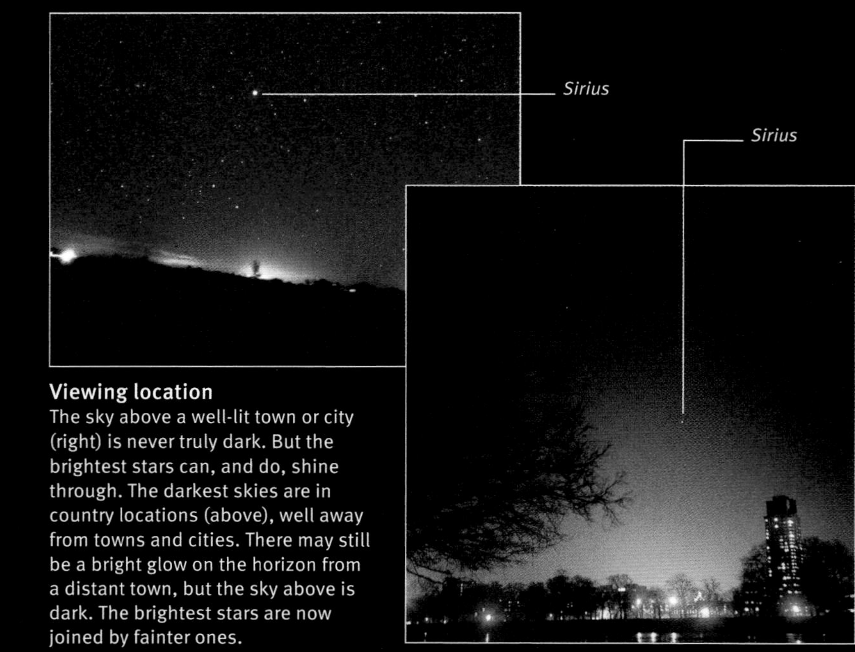

Viewing location
The sky above a well-lit town or city (right) is never truly dark. But the brightest stars can, and do, shine through. The darkest skies are in country locations (above), well away from towns and cities. There may still be a bright glow on the horizon from a distant town, but the sky above is dark. The brightest stars are now joined by fainter ones.

READY TO GO

Any cloud-free night is an observing night, but some are better than others. Use the Monthly Sky Guide (pp.96–121) to find out the phase of the Moon. If the Moon is full, the sky will be flooded with light and stars will be lost from view. Use the Sky Guide and planisphere to plan your viewing.

IDENTIFYING THE STARS
Initially you should aim to identify two or three of the more prominent constellations and the brightest stars. You can build on this base in the nights ahead. Familiarize yourself with the constellation shapes before going outside.

Remember: as Earth spins, our view of a constellation changes as it moves across the sky. And southern-hemisphere observers should know that maps of the Moon and constellations are usually orientated for northern-hemisphere observers.

Once outside, give your eyes time to adjust to the dark. Within ten minutes or so, more stars will appear. Before 30 minutes have passed, your eyes will be fully dark-adapted, and a range of stars of different brightness will be discernible. Use the technique of averted vision to observe fainter objects. Look slightly away rather than directly at the object. It will then appear as an image, formed by the sensitive edge of the eye's retina.

Looking skywards
The darkest part of the sky is the highest part, farthest away from the horizon. If possible, lean against a wall when looking up, since this will keep your body steady.

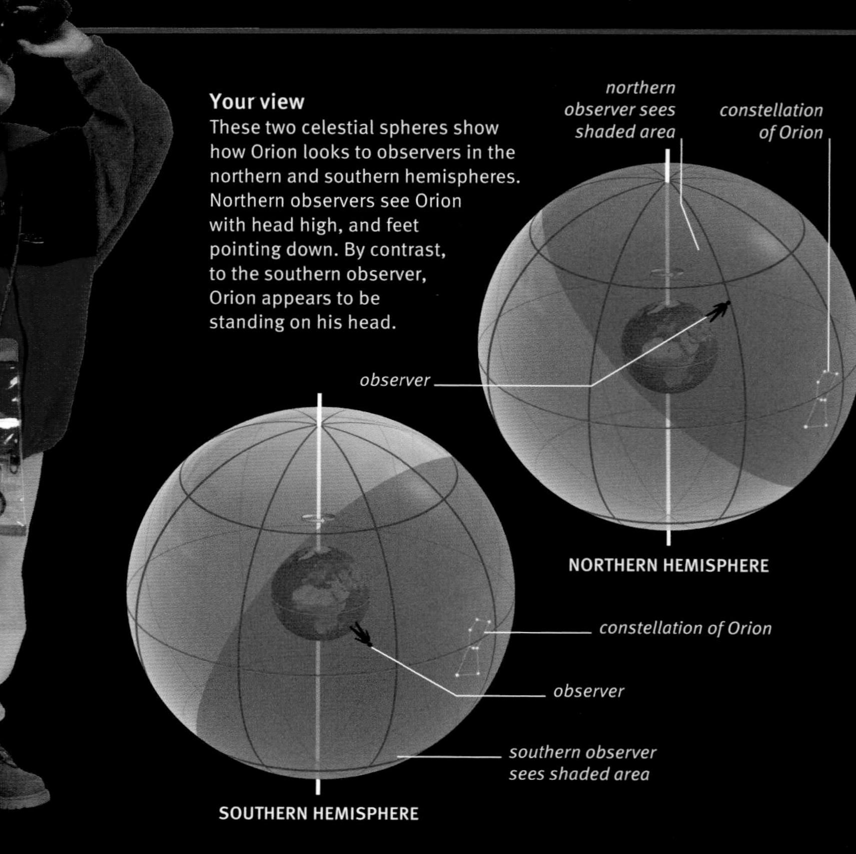

Your view
These two celestial spheres show how Orion looks to observers in the northern and southern hemispheres. Northern observers see Orion with head high, and feet pointing down. By contrast, to the southern observer, Orion appears to be standing on his head.

ENHANCING YOUR VIEW

The naked eye alone can be used to identify the constellations, view planets and meteors, and locate dark and light features on the Moon. However, equipment such as binoculars will enhance your view by collecting more light than the human eye can. They add clarity to your view of the Moon's surface and of stellar objects such as star clusters and the Orion Nebula, as well as revealing objects not visible to the naked eye.

FOCUSING BINOCULARS

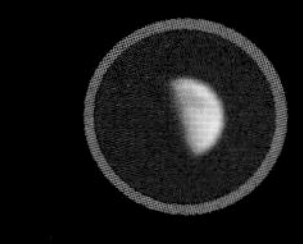

1 Try the binoculars
Everyone's eyesight is different, so you must focus binoculars before use.

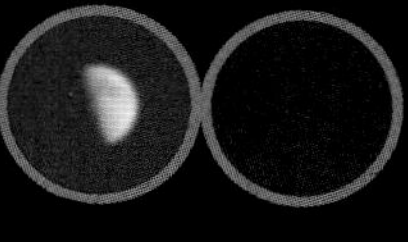

2 Locate eyepiece focus
Find which eyepiece can be focused independently. Look through with that eye closed.

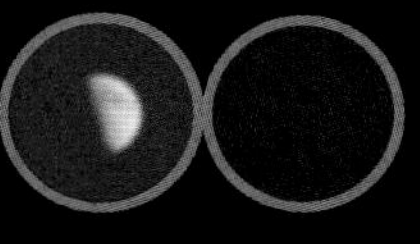

3 Focus left eyepiece
Rotate the binoculars' main, central focusing ring until the left image is in sharp focus.

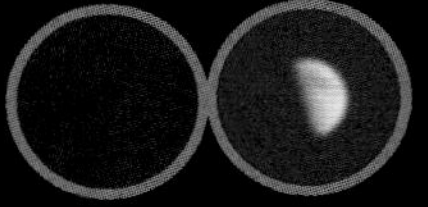

4 Switch eyes, and focus
Open only the other eye. Use the eyepiece focusing ring to bring the image into focus.

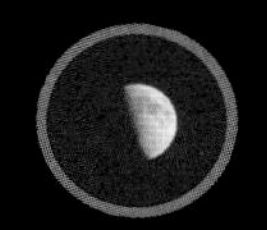

5 Use the binoculars
Both eyepieces should now be in focus. Open both eyes and start observing.

BINOCULARS

The most useful optical aid for a newcomer to astronomy is a pair of binoculars. They are easy to use and portable, and they show the image the right way up (unlike telescopes). They are a combination of two low-powered telescopes, using both eyes for viewing.

Binoculars come in a range of sizes. The two numbers that describe a pair of binoculars are important. The first is magnification; the second, the diameter of the objective lens. Binoculars described as 7 x 50, magnify an object seven times and collect the starlight with a lens 50mm wide. Larger binoculars, which have a lens of 70mm or more and magnifications of 15–20, are used by dedicated astronomers. Binoculars are difficult to keep steady, whatever the size. For a stable image with handheld binoculars, sit down, or lean against a low wall, resting the binoculars on top.

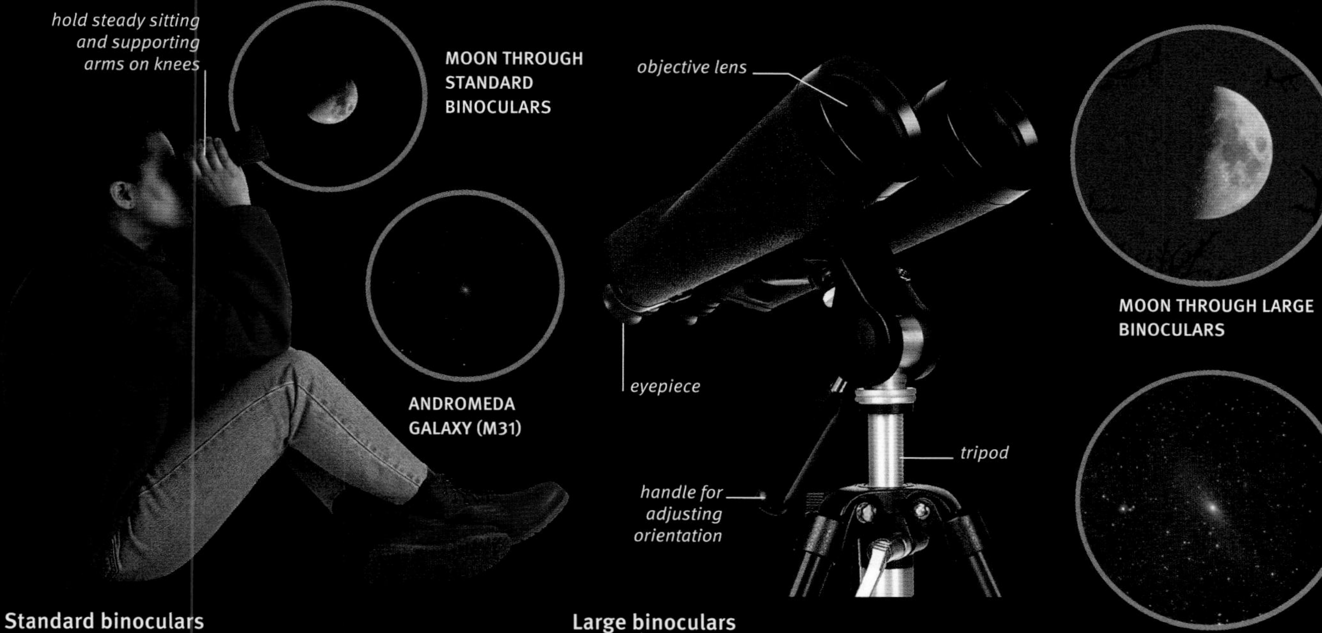

Standard binoculars
Binoculars described as 7 x 50 are ideal for general use. They are light enough to carry around and will reveal about 200 times more stars than can be seen by the naked eye.

Large binoculars
Too heavy to handhold, large binoculars should be supported on a tripod. The larger lens and higher magnification reveal more than standard binoculars.

GETTING EVEN CLOSER

Telescopes bring astronomical objects even closer than binoculars. They gather more light and make faint objects seem brighter and larger.

TELESCOPES

There are two types of telescope: refractors, which use lenses; and reflectors, which use mirrors. Both produce upside-down images, but this is only noticeable when looking at familiar objects. A telescope is further described by the diameter of its main lens or mirror – 75mm (3 inch), for example. The diameter, known as the aperture, is important. Double this, and the light gathered by the telescope quadruples. This light forms an image that is magnified by the eyepiece.

Amateur telescopes are now often controlled by a computer, simplifying the work of locating and tracking particularly faint objects.

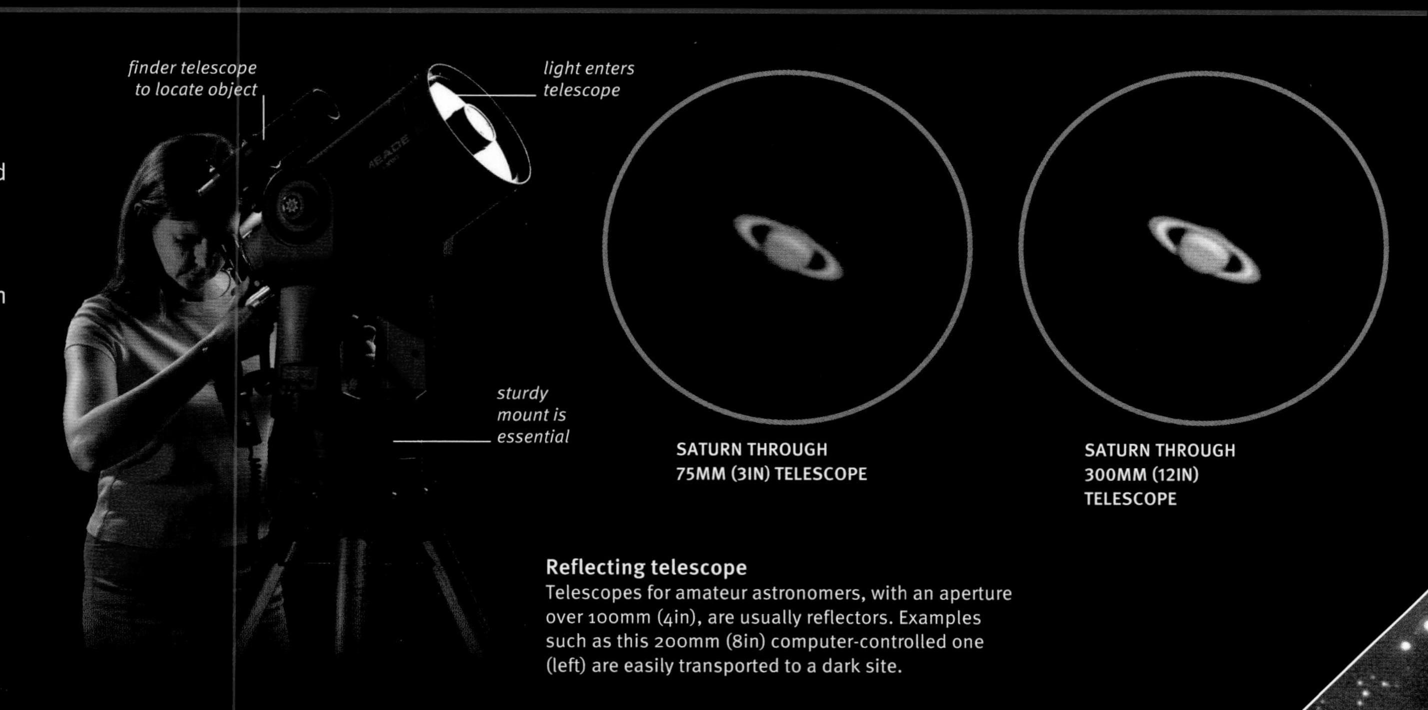

Reflecting telescope
Telescopes for amateur astronomers, with an aperture over 100mm (4in), are usually reflectors. Examples such as this 200mm (8in) computer-controlled one (left) are easily transported to a dark site.

Star hopping

Bright stars and distinctive constellations stand out in the night sky. Patterns such as Orion and the Plough are easily found and, along with stars such as Sirius, can be used as starting points for navigating the sky, hopping from one star to the next. Some well-practised star-hopping routes are shown here. These, and techniques you'll develop with practice, will help you enjoy the night sky.

Ursa Major, the great bear
The third-largest constellation, Ursa Major can be seen from anywhere in Earth's northern hemisphere. Seven of its stars make a saucepan shape called the Plough, or Big Dipper; it is one of the most familiar patterns in the northern sky.

NORTHERN LATITUDES

Ursa Major and Cassiopeia are in the sky all year round. They are on opposite sides of the north celestial pole and make ideal starting points for finding your way around the northern sky.

URSA MAJOR
These routes all start at the Plough, the simple and clear pan-shaped pattern in the tail and rump of the bear.

1 A line from Merak and Dubhe, at the right side of the pan, leads to Polaris. Also known as the Pole Star, Polaris marks the north celestial pole, around which the circumpolar stars revolve. Extend the line for the same distance again to the "W" of stars that is Cassiopeia.

2 Hop from Merak to the bear's front paw; make a second hop of a similar distance to reach Castor, the brightest star in Gemini, the twins.

3 Leo, the unambiguous shape of a crouching lion, is found by extending the line at the left side of the pan. Hop between the back legs of the bear and on to Regulus.

4 Two stars at the end of the bear's tail are the starting point to find Boötes and Virgo. Extend the line away rom the bear to the brilliant star Arcturus. Make a second hop, the same distance again, to find Spica, the brightest star in Virgo.

5 The left end of the pan can also be used to find Lyra. Its bright star, Vega, is about the same distance away but in the opposite direction from the Plough to Leo.

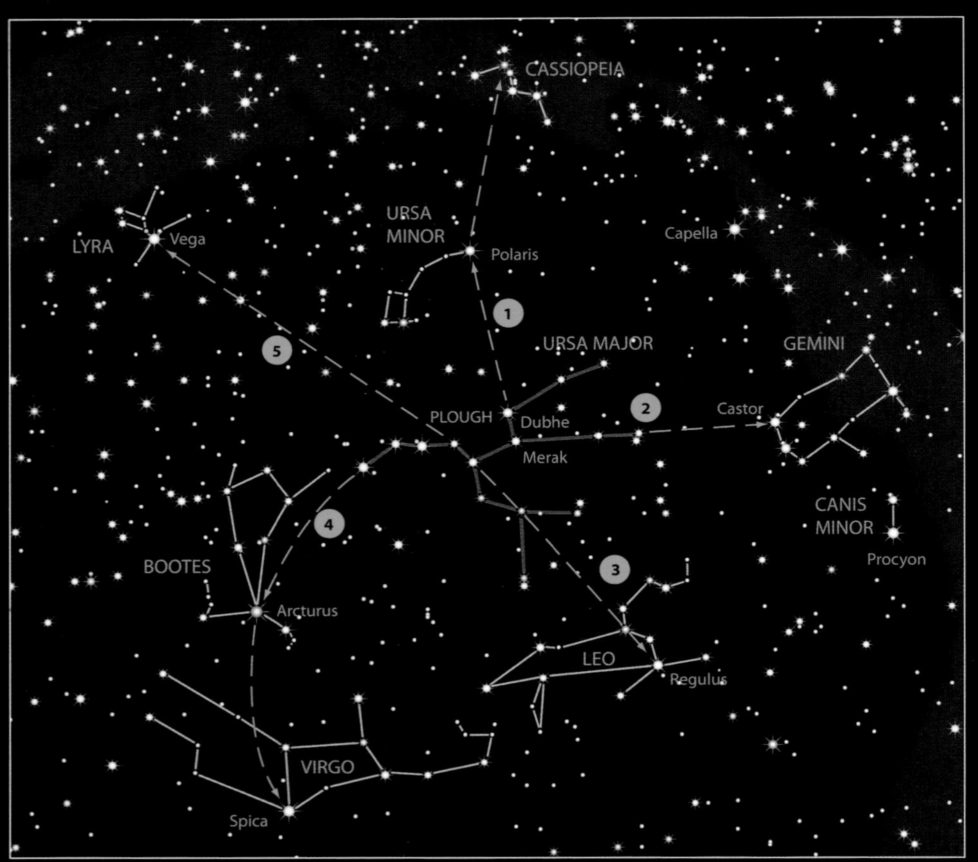

URSA MAJOR

CASSIOPEIA
The "W" shape of Cassiopeia lies within the Milky Way and is a starting point for locating three constellations and one asterism.

1 Hop twice the length of Cassiopeia along the path of the Milky Way. The bright star you find is Deneb, representing the tail of Cygnus, the swan.

2 From Deneb, locate two brighter stars, one at either side of the Milky Way's path. These are Vega (in Lyra) and Altair (in Aquila). The three together make the Summer Triangle.

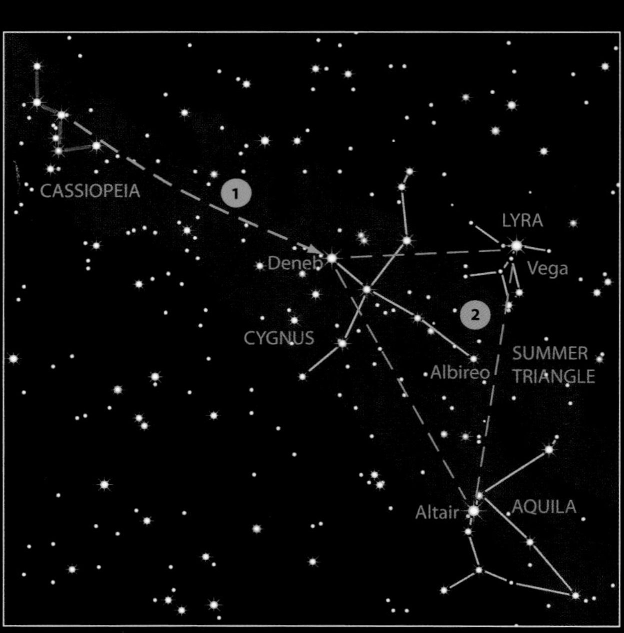

CASSIOPEIA

MID-LATITUDES

The figure of Orion is a prominent presence. Three stars in the hunter's belt are the starting point for exploring a rewarding area of sky.

1 Extend a line from the belt, through Betelgeuse and beyond Orion's raised arm, to reach Gemini, the twins.

2 Hop from the belt of Orion to his head, go the same distance again to the star at the end of Taurus the bull's horn, and carry on to Capella.

3 Hop from the belt to Orion's left hand; hop the same distance again to reach the Pleiades star cluster, passing through Taurus's face on the way.

4 A line from Betelgeuse to Rigel and extended on through a relatively barren area of sky reaches the bright star Achernar.

5 Follow the line of Orion's belt past his lower body and on to the unmistakably brilliant Sirius.

6 Hop from Betelgeuse to Sirius and the same distance again to Procyon to make an equilateral triangle of stars called the Winter Triangle.

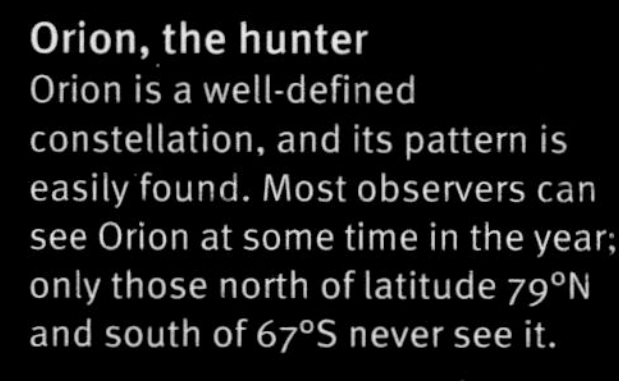

Orion, the hunter
Orion is a well-defined constellation, and its pattern is easily found. Most observers can see Orion at some time in the year; only those north of latitude 79°N and south of 67°S never see it.

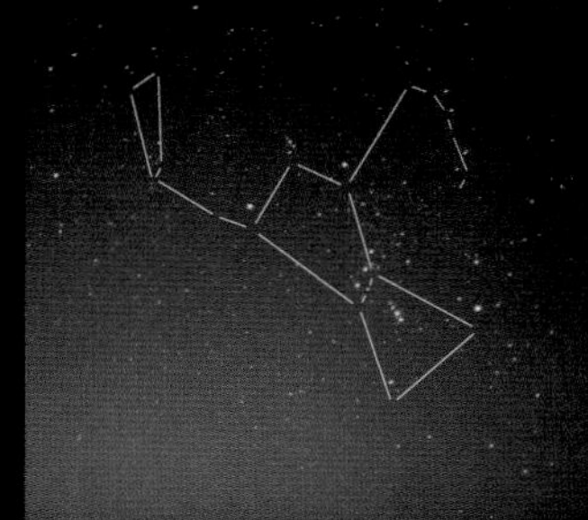

SOUTHERN LATITUDES

The four bright stars in Crux and the two brightest in Centaurus are embedded in the Milky Way's path. Just a hop away are other brilliant stars.

1 Extend the long axis of Crux, the cross, and another line from between the stars Hadar and Rigil Kentaurus, the third-brightest star of all. The two lines meet at the unmarked position of the south celestial pole, which is about halfway to the bright star Achernar.

2 A triangle of lines links the south celestial pole, around which the circumpolar stars revolve, to Achernar, the ninth-brightest star of all, and to Canopus, the second brightest.

3 Find brilliant Sirius, the brightest star in all of the sky, by lengthening and curving the line from Canopus. Hop about the same distance again to Procyon, the eighth-brightest star. Sirius and Procyon are two of the three stars that form the Winter Triangle, shown in the mid-latitude sky map above.

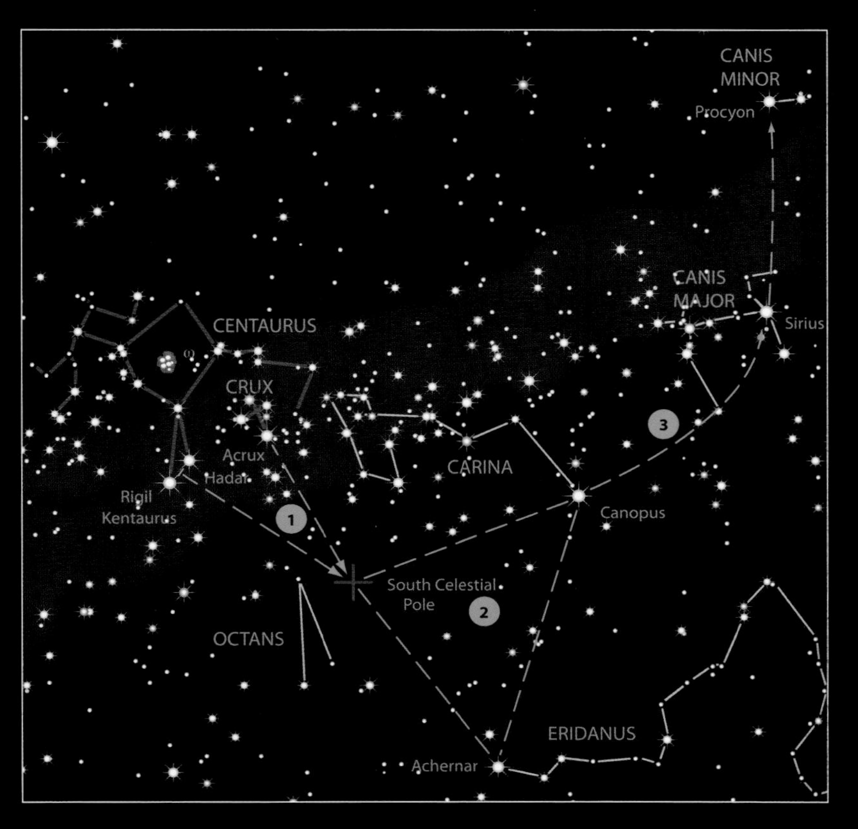

Crux, the southern cross
Crux is the smallest constellation of all, but its four prominent stars and location in the Milky Way make it a favourite of southern observers. It can be seen from anywhere in Earth's southern hemisphere.

The Solar System

The closest celestial objects to Earth are the Moon and the planets. Along with Earth, they are part of the Solar System. The Sun is at the centre of our local space neighbourhood, and the planets, their moons, asteroids, and comets all orbit around it. As all these objects move along their orbits, we are able to observe these fascinating worlds. There are rock spheres covered by craters, volcanoes, and freezing desert; giant planets surrounded by swirling gas; and, once in a while, a dirty-snowball comet makes a spectacular appearance.

Saturn and its rings
Saturn's rings consist of pieces of dirty ice. Here, they have been coloured according to the sizes of their particles, ranging from blue for the smallest, through to green, and then purple for the largest.

What's in the Solar System?

The Solar System consists of the Sun, eight planets, three dwarf planets, more than 160 moons, and billions of asteroids and comets. The Sun is by far the most massive member of the system, lying at its centre; the other objects all orbit around it. They have existed together since they were formed from a cloud of gas and dust about 4.6 billion years ago.

STRUCTURE AND EXTENT

The Sun's immense gravity holds the Solar System together. It pulls on all of the other objects in the system and keeps them on their set paths around the Sun. One complete circuit of a path is called an orbit. Each object in the system – from the smallest asteroid, to Jupiter, the largest planet – orbits the Sun.

The orbits taken by the planets, which are the most significant objects after the Sun, are made close to the plane of the Sun's equator. The asteroids that form the Main Belt, between Mars and Jupiter, also tend to stay close to this plane. Consequently, the planetary part of the Solar System is disc-shaped. The average distance from the Sun to the farthest planet, Neptune, is 4.5 billio km (2.8 billion miles).

Beyond Neptune is the Kuiper Belt – a flattened belt of icy, comet-like bodies. And outside of this are the comets themselves. The comets follow randomly inclined orbits that might be close to the planetary plane, above or below the Sun, or anywhere in between. The comets form the Oort Cloud, which is a huge spherical cloud surrounding the Kuiper Belt and the planetary region. The outer edge of the Oort Cloud, which is some 1.6 light years away and nearly halfway to the closest stars, marks the extent of the Sun's influence. Beyond this lies interstellar space.

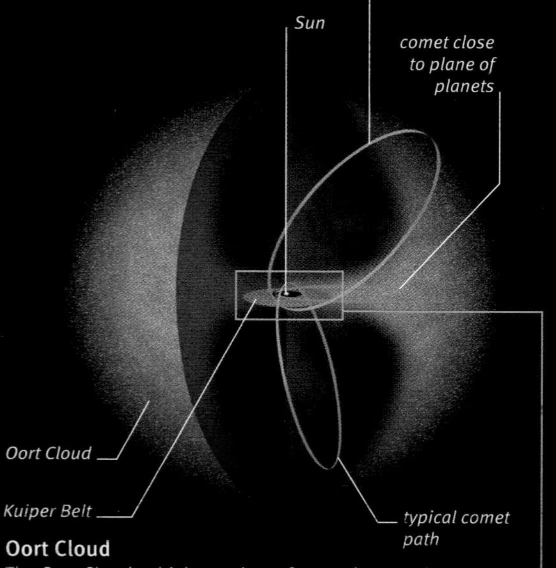

Oort Cloud
The Oort Cloud, which consists of more than a trillion comets, surrounds the planetary region and the Kuiper Belt. Three comet orbits are shown here in green: one extends to the edge of the cloud; a second is close to the plane of the planets; the third is a typical elongated orbit.

Neptune's orbit
Sun
planetary region
Pluto's orbit
Uranus's orbit

Kuiper Belt
Encircling the planetary region, the Kuiper Belt extends from about 6 billion km (3.7 billion miles) from the Sun. It is thought that its outer edge is about 12 billion km (7.4 billion miles) away. The belt is believed to consist of many thousands of objects, more than a thousand of which have been identified.

the largest and most massive planet, Jupiter is the one with the fastest spin

liquid water covers more than 70 per cent of Earth, which is the only planet known to have life

Mars is the most distant rocky planet from the Sun; it is also the coldest

the hottest and slowest spinning planet, Venus is similar in size to Earth

twice as far from the Sun as Saturn is, Uranus is tilted on its side

Mercury is the smallest and fastest-moving planet; it is also the closest to the Sun

Planetary orbits
The planets and asteroids orbit the Sun in an anticlockwise direction when seen from above Earth's North Pole. As they orbit, each one also spins. Orbit size and the time taken to complete an orbit increase with distance from the Sun. Orbits are elliptical, which means that a planet's distance from the Sun varies by many millions of kilometres during its orbit. The planets and their orbits in this image are not shown to scale.

THE PLANETS

The eight Solar System planets all fall into one of two categories: the rocky planets and the gas giants. The rocky planets are the four closest to the Sun – Mercury, Venus, Earth, and Mars. They were formed from rocky and metallic material near the newly born Sun. Each consists of a rocky mantle and crust, surrounding a metal core, yet their surfaces differ widely. Mercury and Venus are hot, lifeless, dry worlds, but while the former is covered in craters, the latter is shrouded in a thick atmosphere that hides a volcanic landscape. By further contrast, Earth is wet and teems with life, and the more distant, colder Mars is a red world of frozen desert.

The gas giants Jupiter, Saturn, Uranus, and Neptune are the four largest planets. They formed from rock, metal, gas, and ice in the cooler outer regions of the disc of material surrounding the young Sun. Each has a rock-rich core surrounded by an unfathomably deep and thick atmosphere. All four have a ring system encircling them and a large number of orbiting moons.

Rocky landscape
This panoramic view of Mars was produced from 28 images taken in February 2006 by the rover Opportunity during its exploration of the planet. The view is of the western edge of Erebus Crater. Layered rocks can be seen in the 1m- (3ft-) thick crater wall.

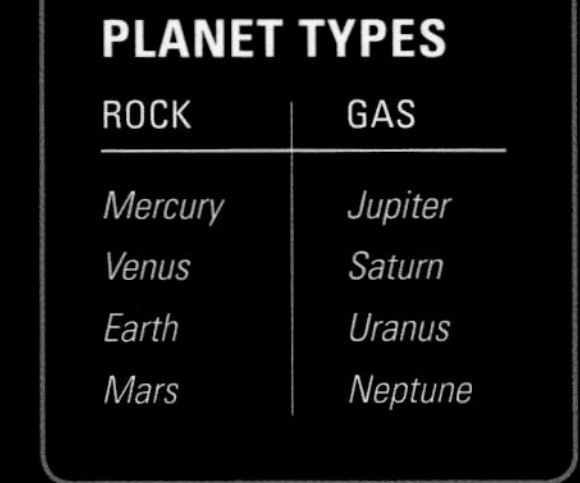

PLANET TYPES

ROCK	GAS
Mercury	*Jupiter*
Venus	*Saturn*
Earth	*Uranus*
Mars	*Neptune*

Gaseous atmosphere
The gas giants do not have solid surfaces; what we see is the cloud-top layer of their atmospheres. Jupiter's visible surface shows bands of swirling gas – and spots, which are giant storms.

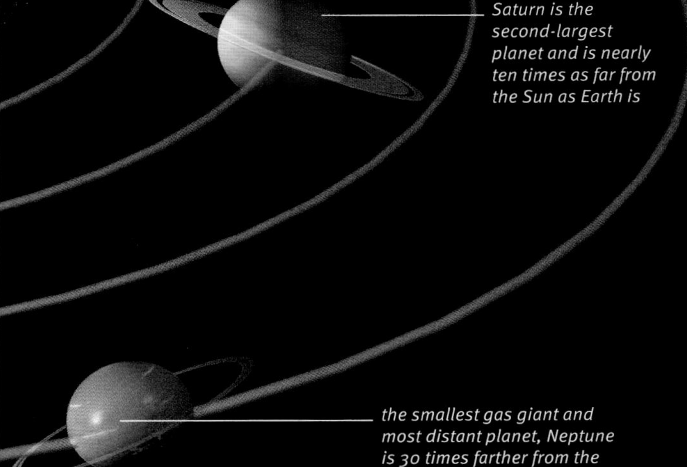

MOONS AND RINGS

At present, 162 planetary moons are known to exist, but more are likely to be found as observing techniques improve. The majority belong to the gas giants; only Mercury and Venus are moonless. The moons are made of either rock or a mix of rock and ice, and they orbit around their planets like mini solar systems. The largest is Jupiter's moon Ganymede, which is bigger than Mercury. The smallest are hill-sized irregular lumps. All four giant planets also have ring systems comprising pieces of dirty ice. The most extensive of these is the one around Saturn.

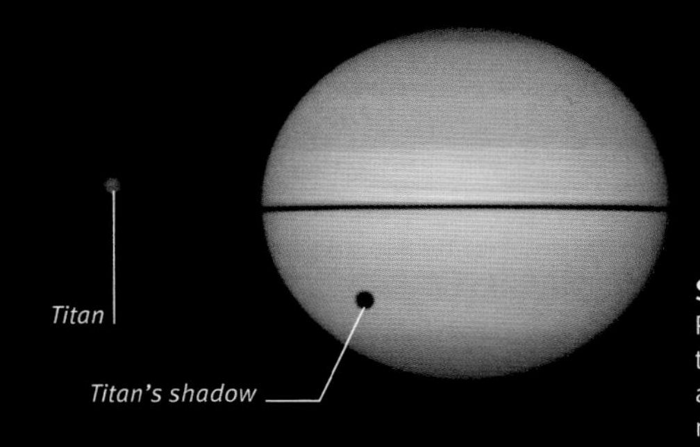

Saturn's rings
Paper thin compared to their width, Saturn's rings are barely visible here. Four moons, including Mimas, are clustered to the right of the rings.

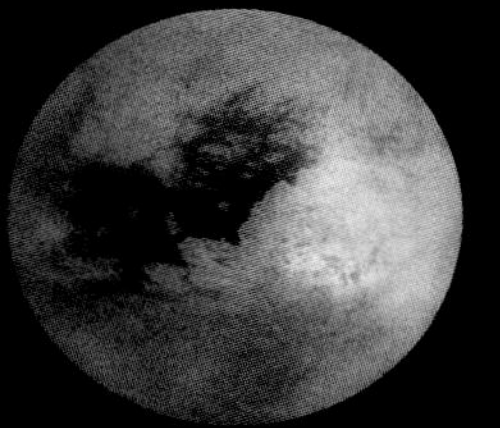
TITAN

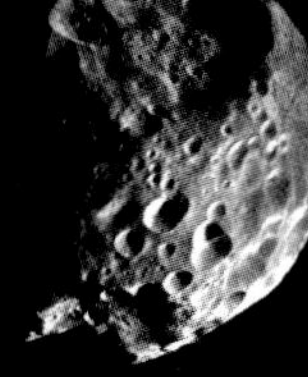
PHOEBE

Contrasting moons
Titan is the second-largest moon in the Solar System, and the only one with a substantial atmosphere. A nitrogen-rich smoggy haze envelops the moon, but here it has been removed to show the moon's surface. Phoebe, another of Saturn's moons, is more typical: it is 230km (143 miles) long, potato-shaped, and covered in craters.

MINOR MEMBERS

There are more than a trillion smaller bodies in the Solar System. The majority are comets and asteroids, followed by thousands of Kuiper Belt objects, and finally the dwarf planets: Eris and Pluto, two icy rock bodies beyond Neptune; and rocky Ceres in the Main Belt of asteroids. Over a billion asteroids are thought to exist. More than 200,000 are in the Main Belt between Mars and Jupiter; these are the remnants of a failed process of planet formation. Most of the asteroids are irregular-shaped lumps of rock. Comets are huge, dirty snowballs within the Oort Cloud. When one leaves the cloud and travels in towards the Sun, it becomes large enough and bright enough to be seen.

Comet Hale–Bopp
Hale–Bopp was seen in 1997 as it travelled close to the Sun. Its large head and its tails of gas (blue) and dust (white) made it one of the brightest 20th-century comets.

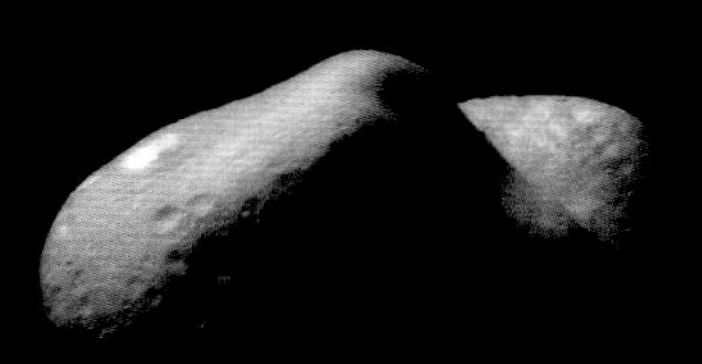

Asteroid Eros
Orbiting outside the Main Belt between Mars and Earth, the potato-shaped Eros is 31km (19 miles) long. Its surface is covered in craters where other bodies have struck it.

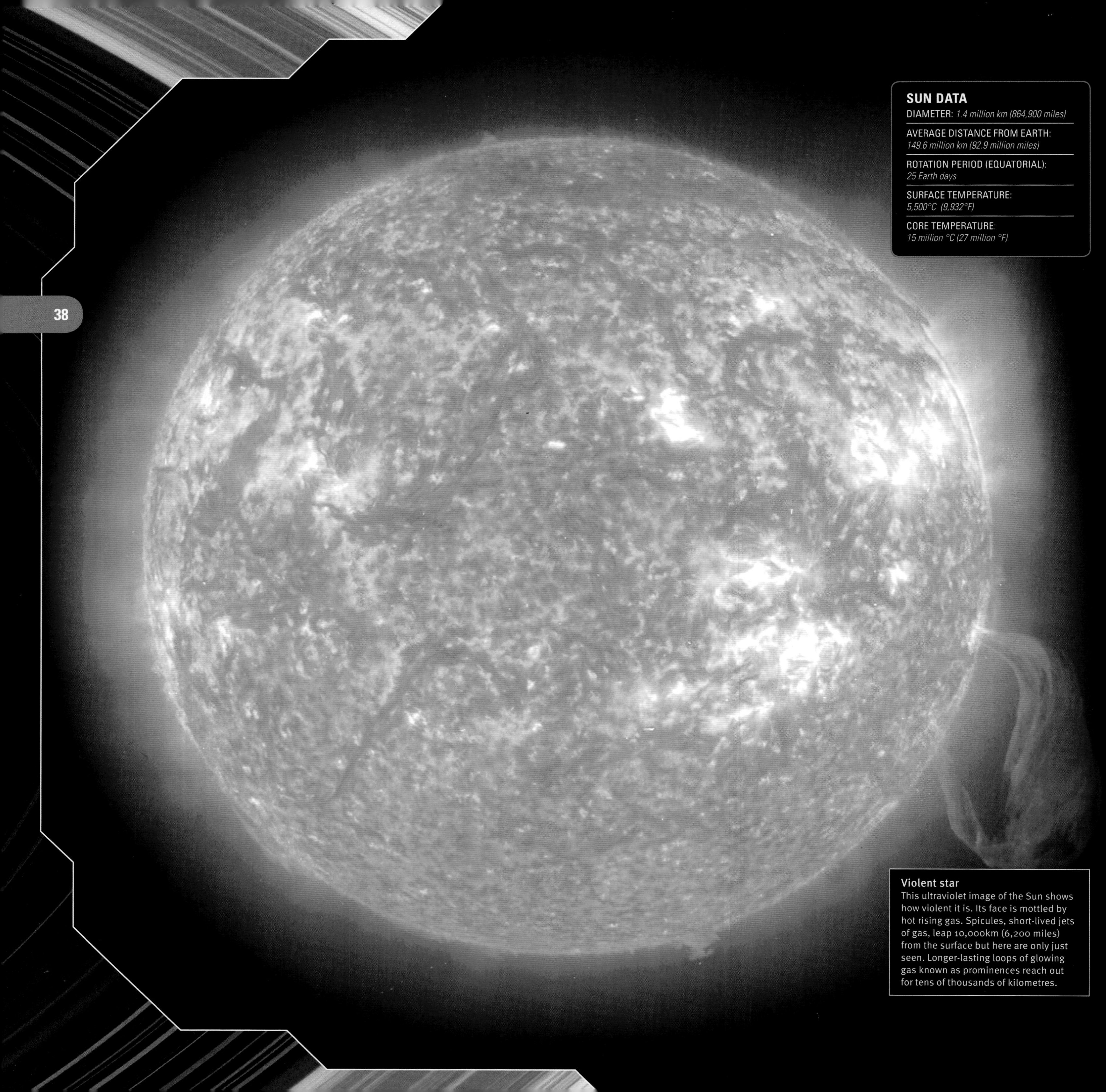

SUN DATA

DIAMETER: *1.4 million km (864,900 miles)*

AVERAGE DISTANCE FROM EARTH: *149.6 million km (92.9 million miles)*

ROTATION PERIOD (EQUATORIAL): *25 Earth days*

SURFACE TEMPERATURE: *5,500°C (9,932°F)*

CORE TEMPERATURE: *15 million °C (27 million °F)*

Violent star
This ultraviolet image of the Sun shows how violent it is. Its face is mottled by hot rising gas. Spicules, short-lived jets of gas, leap 10,000km (6,200 miles) from the surface but here are only just seen. Longer-lasting loops of glowing gas known as prominences reach out for tens of thousands of kilometres.

The Sun

The Sun is the closest star to Earth. Like all other stars, it is a vast ball of incredibly hot, brilliant gas. Gravity keeps the Sun together by pulling the gas in towards the centre, where it converts hydrogen to helium and in the process produces heat and light. The Sun has been shining for about 4.6 billion years and will do so for about another 5 billion.

FEATURES

The Sun is about three-quarters hydrogen and a quarter helium, with small amounts of 90 or so other elements. About 60 per cent of all this is in the core, where the temperature and pressure are extremely high and nuclear reactions occur.

The Sun is not solid but has a visible surface, the photosphere. The temperature of the photosphere gives the Sun its colour. It consists of constantly renewing granules of rising gas, each of which is about 1,000km (620 miles) across. Beyond the photosphere is the Sun's atmosphere, which is not normally visible. Nearest the Sun is the chromosphere, which extends out about 5,000km (3,100 miles), and outside of this is the corona, which extends for millions of kilometres into space.

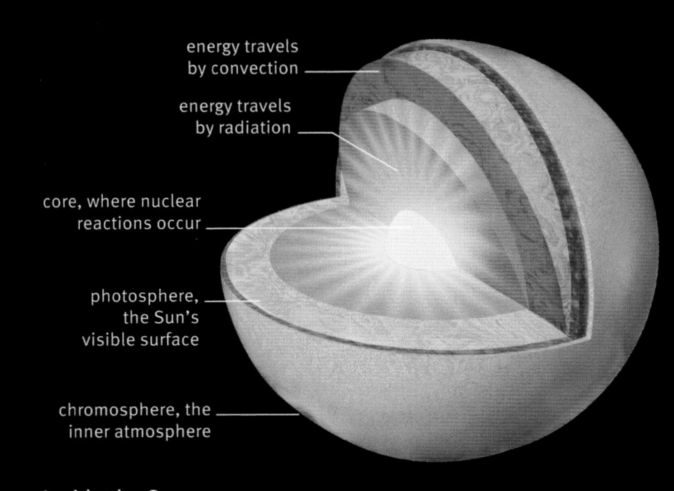

Inside the Sun
The Sun converts hydrogen to helium at about 600 million tons a second. Energy produced moves through the Sun by radiation, then nearer the surface by convection, and is released via the photosphere.

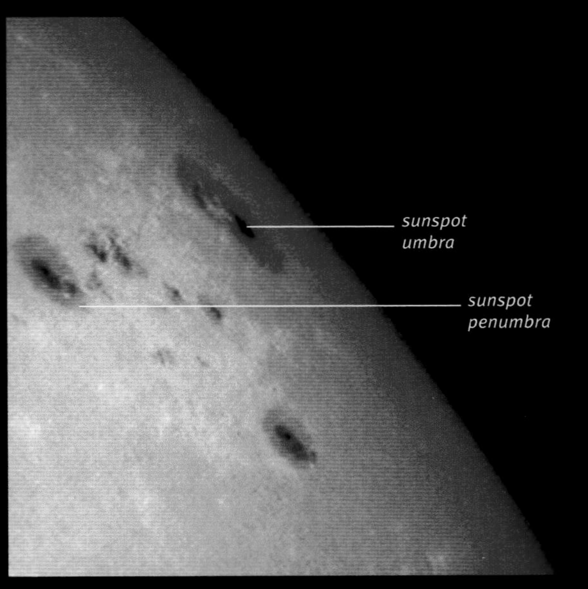

Sunspots
Dark patches on the Sun's surface are called sunspots. These are relatively cool regions of the photosphere – about 1,500°C (2,700°F) cooler than the surrounding surface. Sunspots appear periodically, usually in pairs or groups, between 40 degrees north and 40 degrees south of the Sun's equator. They are hundreds of thousands of kilometres wide and last for a few weeks.

OBSERVING THE SUN

The Sun cannot be observed directly, but it is possible to view its disc safely. One way to do this is by projecting the Sun's image onto a white card using binoculars or a telescope.

A total eclipse of the Sun is a dramatic event that offers the opportunity to see the Sun's outer atmosphere and prominences flaring from the surface. Spectacular light displays called aurorae (singularly aurora) can be seen by observers in the most northerly and southerly latitudes. These are produced by the interaction between particles from the Sun and Earth's upper atmosphere.

WARNING

Never look at the Sun directly with the naked eye or with any instrument. The light will burn your retina, causing permanent blindness.

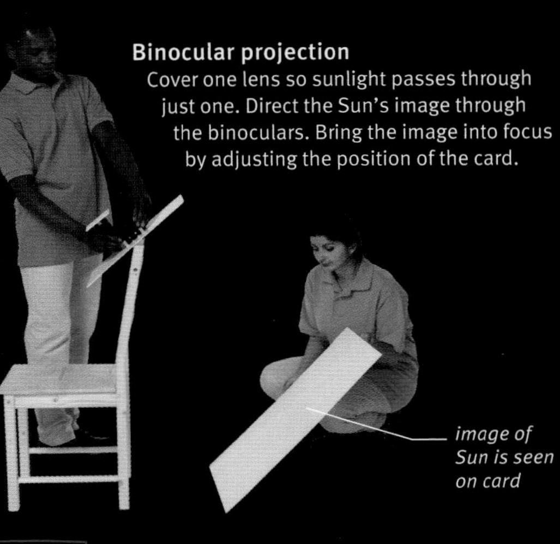

Binocular projection
Cover one lens so sunlight passes through just one. Direct the Sun's image through the binoculars. Bring the image into focus by adjusting the position of the card.

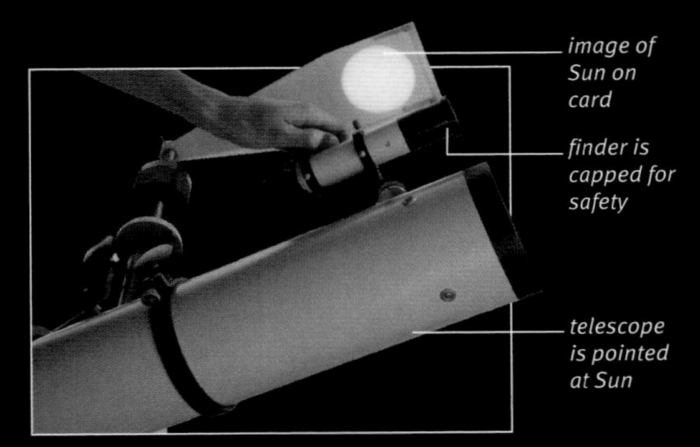

Telescope projection
Position the card about 50cm (20in) from the telescope's eyepiece, and aim the telescope at the Sun. To sharpen the Sun's image, adjust the eyepiece. Any sunspots will be seen as blackened dots on the Sun's disc.

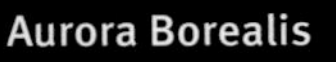

Aurora Borealis
Also known as the Northern Lights, the Aurora Borealis can be seen from locations north of about latitude 50°N. The colourful display of glowing gas shown above was captured just after dusk from the Yukon Territory, Canada. The Aurora Australis (Southern Lights) can be seen south of about 50°S.

Total eclipse
The Sun and Moon appear the same size in the sky, so when the Moon passes directly in front of the Sun, the latter's disc is completely hidden. The Sun's outermost layer, its corona, is then visible. In this photo, red prominences are also seen.

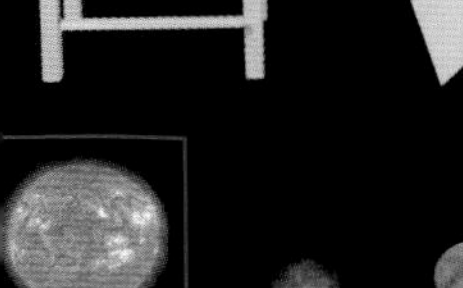
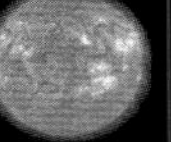

PLANET DATA

DIAMETER *3,476km (2,160 miles)*

AVERAGE DISTANCE FROM EARTH *384,400km (238,900 miles)*

ROTATION PERIOD *27.3 Earth days*

SURFACE TEMPERATURE *–150 to +120°C (–240 to +250°F)*

Unseen Moon
This view of the Moon cannot be seen from Earth; the Earth-facing side is at lower left. The impacts that produced the Moon's large craters cracked the crust so badly that lava flooded the crater floors. It then solidified, producing the darker areas.

The Moon

he Moon dominates the night sky. It is our only natural satellite and closest space eighbour. Many think this cold, dry, lifeless ball of rock was formed when a Mars-ized asteroid collided with Earth about 4.5 billion years ago. Molten material from arth and the asteroid formed the Moon, before cooling and solidifying.

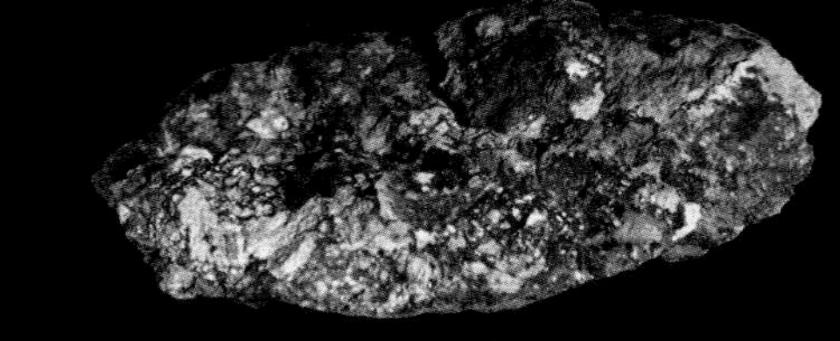

FEATURES

ne lunar crust is solid and rigid. It is about 8km (30 miles) thick on the Earth-facing side, nd 74km (46 miles) thick on the far side. Below a rocky mantle, and because the interior ets hotter with depth, this is partly molten. small iron core may be at the Moon's centre. ombardments by asteroids and meteorites ave pulverized the surface, producing a lunar oil, the regolith, some 5–10m (16–33ft) thick.

Boulders that have been blasted out of both ear and distant craters litter the landscape. The raters, formed by impacting asteroids, range om bowl-shaped ones, less than 10km (6 miles) cross, to features more than 150km (90 miles) ide. These larger ones have been flooded by ava that seeped from inside the Moon.

Lunar landscape
The Moon's mountains are simply the walls of its huge craters and can be up to 5km (3 miles) high. Below, astronaut Harrison Schmitt stands on the rim of the 110m- (356ft-) wide Shorty Crater. Left, an astronaut is dwarfed by a boulder from an impact crater.

Moon rock
Between 1969 and 1972, 12 astronauts walked on the Moon. They returned with more than 2,000 samples of rock, pebbles and dust, and core material. They are very old and mainly consist of silicate rocks and lava from volcanoes.

THE VIEW FROM EARTH

he Moon appears to be the largest object in the ight sky. It produces no light of its own but shines y reflecting light from the Sun. From Earth we nly ever see one side of the Moon because the Aoon spins around in the same amount of time s it takes to complete one orbit around Earth.

The Moon seems to be a different shape rom day to day. But these shapes, known s phases, are simply the different amounts f the Moon's sunlit side visible from Earth. complete cycle of phases lasts 29.5 days.

Phases of the Moon
The Moon's phase changes throughout the month. Two days after the new Moon, a thin crescent appears in the evening sky. A week after the new Moon, half the disc is visible. A week later, the Moon is full, as the whole bright Earth-facing hemisphere is lit by the Sun.

sunlight

NEW MOON

the face seen from Earth is unlit and cannot be seen clearly from Earth

CRESCENT

a thin crescent, seen after sunset, is often called the new Moon

FIRST QUARTER

the Moon is waxing (growing) and has completed a quarter of its orbit

WAXING GIBBOUS

about three-quarters of the sunlit side is seen. Gibbous means lump-shaped

FULL MOON

the Moon is on the opposite side of Earth to the Sun, and we see a fully lit near side

WANING GIBBOUS

the Moon has completed more than half its cycle and is waning (shrinking)

LAST QUARTER

only a quarter of the phase cycle remains; half the near side is lit

CRESCENT

the last sliver of sunlit Moon is visible as the phase cycle ends

Orbital path

Earth

NORTHERN HEMISPHERE

SOUTHERN HEMISPHERE

Changing views
The European astronomers who first mapped the Moon called he top "north" and the bottom "south". But southern-emisphere observers see the Moon's south pole at the top

Daytime Moon
When the Moon is at quarter phase, it is often visible in the daytime. As the Sun is setting, the first-quartered Moon is in the south. It appears much less bright because of the daylight, but surface features can still be seen

Observing the Moon

The Moon's surface is dry, dusty, dark, and dead. The lunar landscape has remained virtually unchanged for millions of years. A "new" crater, wider than 1km (0.6 miles) across, is formed on the Earth-facing side on average every 40,000 years. A glance at the Moon will reveal that it has two types of terrain: large dark plains, called maria (mare in the singular); and brighter, heavily cratered highland regions.

THE NEAR SIDE

Very early in the Moon's life, when it was much closer to Earth and the interior was hotter, it became locked such that one face – the near side – always pointed towards Earth. This affected the crust, and not only is the near side, on average, 5km (3 miles) lower than the far side, the near-side crust is about 25km (15 miles) thinner. Deep craters on the near side were filled with volcanic lava in their early history; this did not happen on the far side. Dark lava plains now cover about half the near side. When the Moon was first mapped, astronomers thought these dark regions were water and referred to each as a mare (sea) or an oceanus (ocean). It is useful to learn their names and positions. More recently formed craters can be seen within the dark flat maria. The higher mountainous regions, which are older and have more craters, are twice as bright as the maria.

Naked-eye view
Dark and light features are visible. The full Moon looks like a face, with Mare Imbrium and Mare Serenitatis as eyes, and Mare Nubium and Mare Cognitum as the mouth.

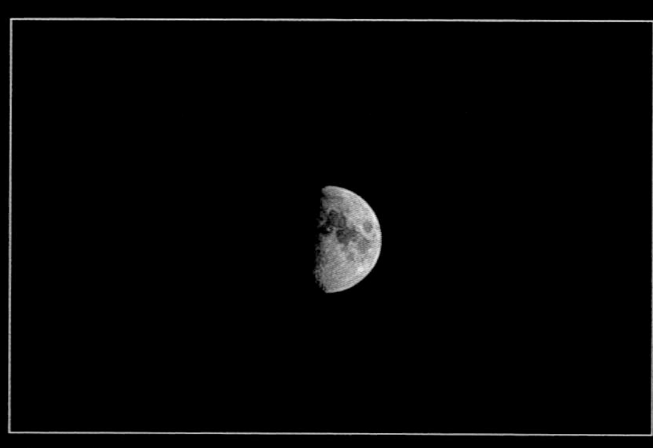

Binocular view
The Moon is still seen as a whole but a well-supported pair of binoculars reveal surface features, such as large craters. The uneven nature of the terminator (see opposite) is also visible.

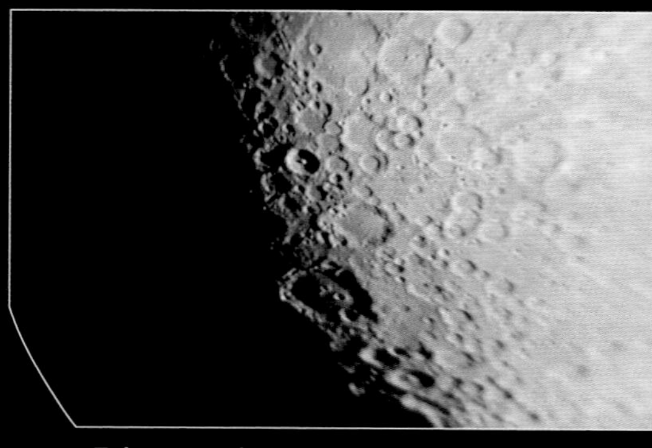

Telescope view
Now only a part of the Moon is visible. Thousands of smaller craters and details of shadows, mountains, and valleys become distinct. The usable magnification depends on the turbulence of Earth's atmosphere.

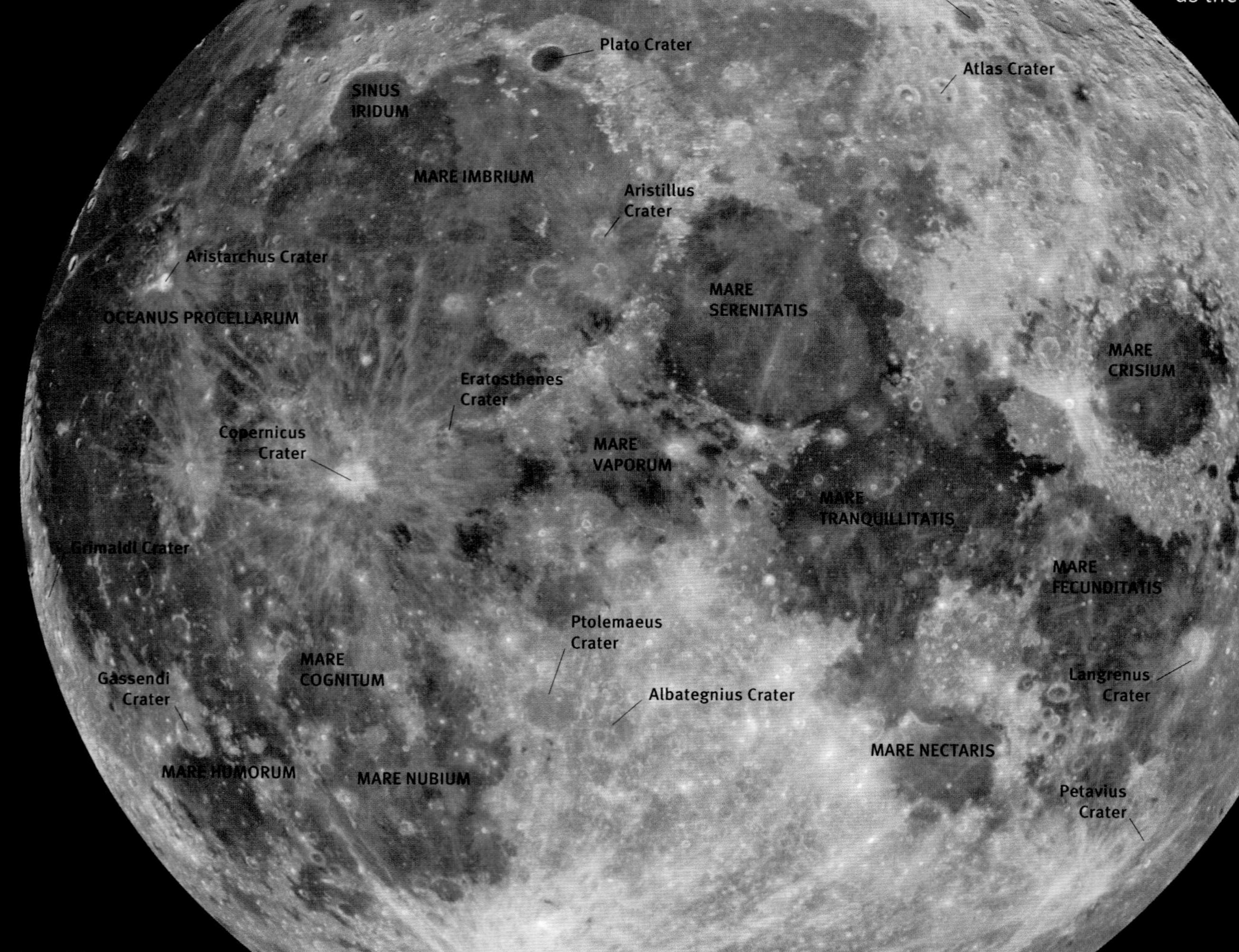

Near-side view
The near side is dominated by the large dark Oceanus Procellarum on the western side. Bright spots in the region are more recently formed craters. Copernicus is particularly prominent.

SURFACE FEATURES

You can train your eye to observe detail by concentrating on small parts of the lunar surface and sketching the craters. You will soon realize that there are far more small craters than large ones. The biggest visible crater is Mare Imbrium, some 1,100km (680 miles) across, the formation of which almost broke the Moon apart. About 4 per cent of the craters are not circular. These were produced by impactors arriving at very low angles. Some craters are "ghosts" – only the peaks of the walls protrude above the mare lava. The lunar surface also has some interesting valleys. These were not caused by flowing water but are mainly the remnants of lava tubes that have emptied and then caved in. Smaller valleys are called rills. Craters such as Gassendi and Hevelius have many rills criss-crossing their lava-filled basins. These were formed when the lava cooled and contracted.

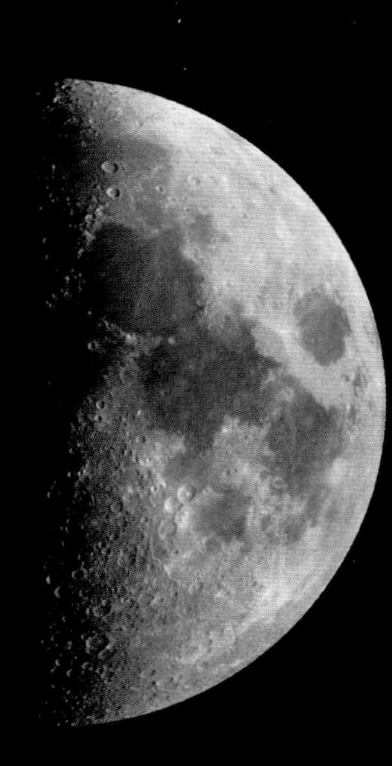

Terminator
The terminator is the boundary between the sunlit part of the Moon and the dark part. It moves around the Moon during the month. Shadows at the terminator are very long, and features such as mountains are thrown into relief.

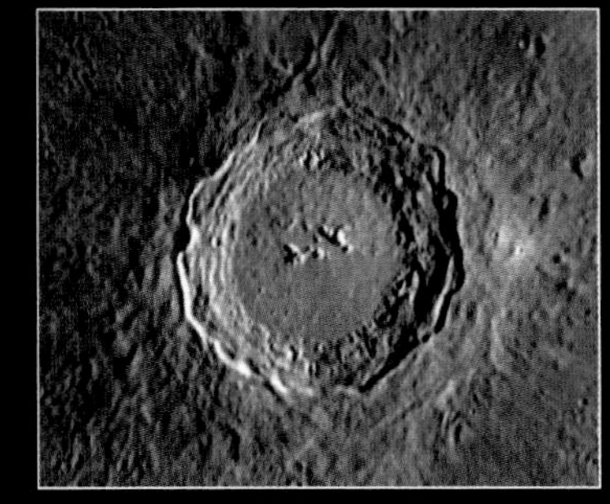

NOON: THE SUN IS OVERHEAD

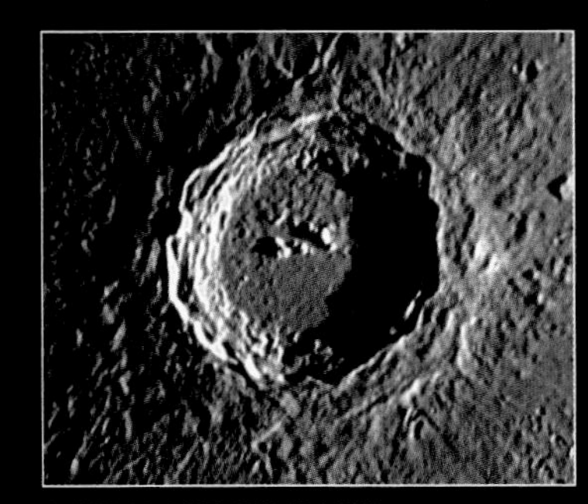

SUNRISE: THE SUN IS LOW

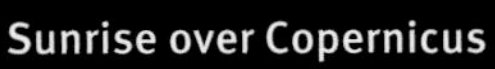

Sunrise over Copernicus
This crater is 800 million years old, and is 91km (57 miles) wide and 3.7km (2.3 miles) deep. It looks "washed out" when the Sun is overhead (top), but at sunrise (above), the long shadows highlight the dramatic nature of the crater's collapsed inner walls and the central mountain. The surrounding hills are also more visible.

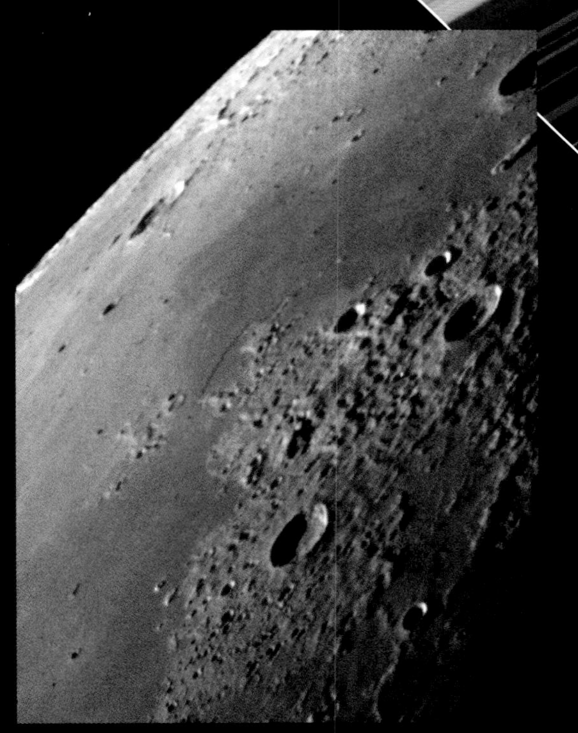

Sinus Iridum
The beautiful Sinus Iridum (Bay of Rainbows) is the western half of a 260km- (160-mile-) diameter impact crater that tilted over when the central regions of Mare Imbrium subsided. The eastern side of the crater was then lost from view as lava from beneath the lunar crust completely flooded the gigantic mare.

OBSERVING CRATERS

Craters appear at their most dramatic when they are near the terminator, which sweeps across the lunar surface as the lunar month progresses. The craters listed here have been divided into four groups according to when they are close to the terminator – that is, whether they are best observed when the Moon is a new waxing crescent, at first quarter, at last quarter, or an old waning crescent. The craters can all be viewed with binoculars. When the Moon is full, concentrate on the cratered regions around its rim.

BEST OBSERVATION TIME

Waxing crescent	*Last quarter*
First quarter	*Waning crescent*
Endymion	*Plato*
Langrenus	*Tycho*
Petavius	*Eratosthenes*
Atlas	*Aristarchus*
Albategnius	*Gassendi*
Ptolemaeus	*Grimaldi*
Aristillus	*Pythagoras*

Tycho Crater
Visible in the centre of this image, the Tycho Crater is an impressive sight through binoculars. This "young", 85km- (52-mile-) wide crater was formed 100 million years ago. In its centre is a 3km- (1.8-mile-) high mountain peak, formed when the underlying rock relaxed after being depressed by the pressure of the impact. Bright rays of material that were ejected from the crater can be seen splashed over the nearby lunar surface.

LUNAR ECLIPSES

Total lunar eclipses occur only when the Moon is full and the Sun, Earth, and Moon are lined up (see p.27). Such eclipses can be seen from anywhere on the night-time hemisphere of Earth, and no special equipment is needed to observe one. During the eclipse, the Moon is in the Earth's shadow, and the Earth is preventing sunlight from reaching its surface. At totality, a small amount of indirect sunlight still manages to reach the Moon. This travels close to the Earth and through its atmosphere. As a result, the light is a deep red-orange colour.

Onset of eclipse
The lower edge is totally eclipsed. Light from part of the Sun continues to shine on the rest of the Moon.

Halfway stage
About half of the Moon is now totally eclipsed, while the other half is still illuminated by sunlight.

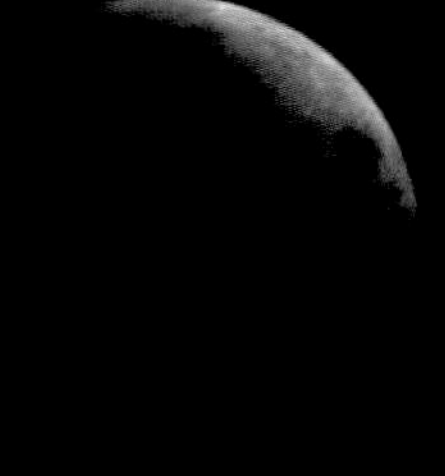

Darkened crescent
Most of the Moon is in Earth's shadow. The crescent is faint because it is in the penumbra.

Totality
The Moon is completely inside Earth's shadow and is a dusky reddish colour.

Mercury

Rocky, crater-covered Mercury is the smallest planet and the closest to the Sun. It has the barest of atmospheres, and conditions on its surface are extremely harsh. Although visible to the naked eye, Mercury is never far from the Sun and so is often difficult to find.

PLANET DATA

DIAMETER *4,875km (3,029 miles)*

AVERAGE DISTANCE FROM SUN *57.9 million km (36 million miles)*

ORBITAL PERIOD *88 Earth days*

ROTATION PERIOD *59 Earth days*

SURFACE TEMPERATURE *–180°C to +430°C (–292°F to +806°F)*

NUMBER OF MOONS *none*

Cratered world
Mercury is a dry, rocky world reminiscent of Earth's Moon. It has a very old surface, much of which is very heavily cratered rather like the lunar highland. The remaining area is younger; it consists of more lightly cratered plains of solidified volcanic lava, rather similar to the lunar maria.

FEATURES

Mercury is very dense compared to the other rocky planets, and beneath its rocky crust and mantle is a large iron core. It is not certain whether this is because the region of the Solar System where Mercury was formed was rich in iron, or whether the early cratering of Mercury has eroded its rocky mantle. The core formed when the heavy iron sank within the young planet.

The planet's surface is pockmarked by thousands of craters – from small, bowl-shaped ones to the vast Caloris Basin (right). Meteorite bombardment churned up the surface and produced a powdery soil that reflects little light and is very dark, just like lunar soil.

Mercury is too small and too hot to have anything more than a very thin atmosphere. This is either captured from the solar wind of gas constantly escaping the Sun or produced by the roasting of surface rocks. Much escapes in the daytime, and it is constantly replenished.

Mercury has the most elongated planetary orbit: its distance from the Sun varies between 46 million km (28.6 million miles) and 69.8 million km (43.4 million miles), and it spins round exactly three times during every two orbits. This makes a "day" on Mercury last 176 Earth days.

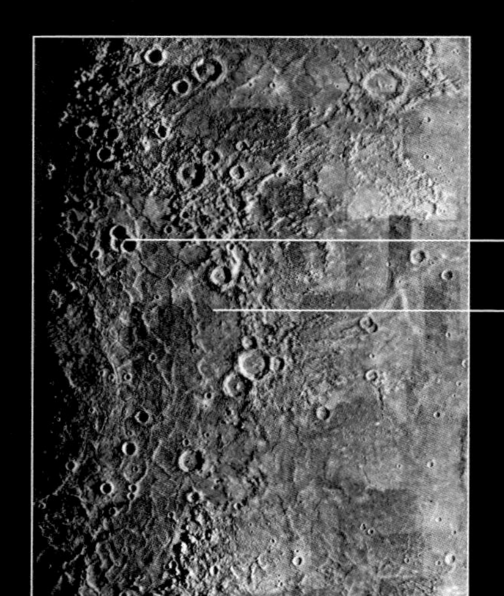

Caloris Basin
The huge, multiringed Caloris Basin impact crater is about 1,350km (840 miles) across (its centre lies to the left of this image). The asteroid that caused it was about 100km (60 miles) wide. The resulting seismic waves traversed Mercury, shattering the surface on the other side, before travelling back to cause fracturing and landslides.

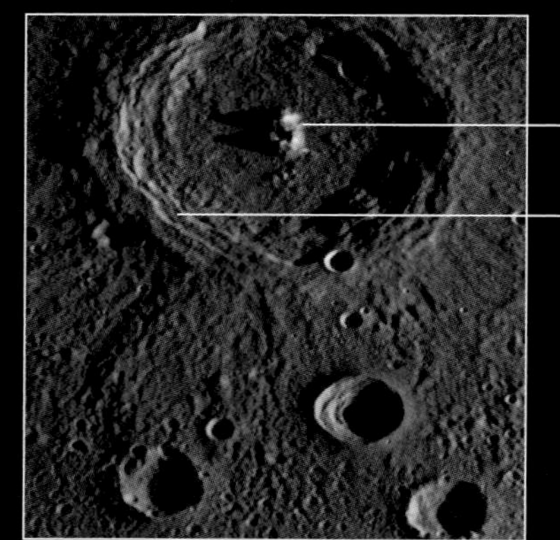

Brahms Crater
This bowl-shaped crater is 97km (60 miles) across and was formed when an asteroid hit Mercury about 3.5 billion years ago. It has a prominent central mountainous peak and inner walls that have slipped inwards due to structural weakness and the influence of gravity.

OBSERVING THE TWO PLANETS

Mercury and Venus are both inside Earth's orbit, so they never appear far from the Sun. Mercury is seen closer to the Sun than Venus, which orbits at a greater distance. As a result, the two are only seen in the early morning eastern sky before the Sun rises, and in the western evening sky after the Sun has set. Mercury is only visible for around a fortnight when the planet is at elongation. Venus can be very prominent at elongation and is then often by far the brightest body in the sky. Dates of elongation are included in the Special Events listings of the Monthly Sky Guide (pp.96–121).

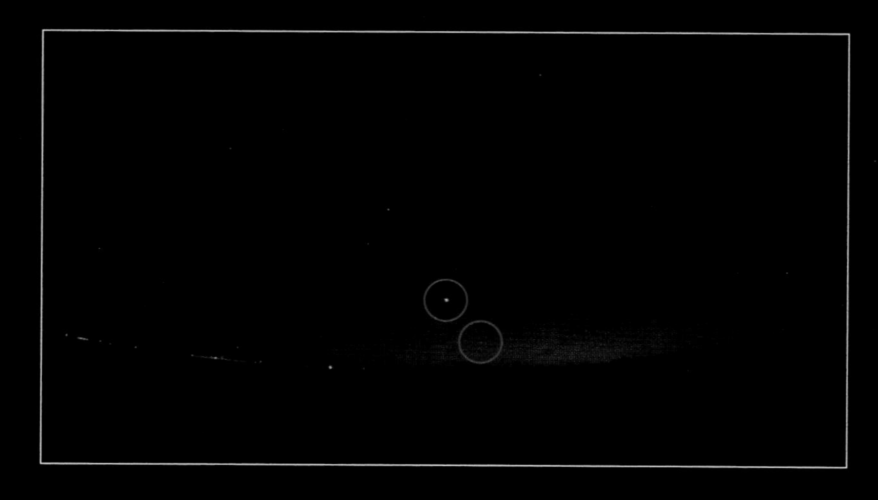

Naked-eye view
Here, the Sun is below the horizon and is illuminating the sky near the horizon. The brighter of the circled planets is Venus. Mercury is fainter because it is both smaller and has a rocky surface that reflects less sunlight. Few stars are seen in the vicinity because their light has been absorbed by Earth's atmosphere. Both planets go through a phase cycle similar to the Moon's, but these are only seen with optical aid.

Venus

Venus is the second rocky planet from the Sun and the brightest in Earth's sky. It is Earth's inner neighbour and twin, with a similar size and mass. However, Venus has lost all of its water, and its carbon dioxide has escaped to form a very dense atmosphere that acts like a greenhouse, trapping in heat.

FEATURES

The similarity between Venus and Earth means that, like Earth, Venus has a hot rocky mantle beneath its solid surface crust. This is the source of Venus's spasmodic volcanic activity. Below this is an iron-nickel core, which has a central solid region surrounded by a liquid-metal outer region. Due to the baking-hot temperature at Venus's surface, the vast majority of the water that was released from the mantle rocks in the past has escaped into space. As a result, the dry mantle rocks are very viscous, and Venus has no moving plates and no mountain ranges.

Radar systems on board orbiting spacecraft have penetrated Venus's clouds and mapped its surface. It is dominated by volcanic features: 85 per cent of it is covered with low-lying plains of lava produced about 500 million years ago, and hundreds of volcanoes are visible. Large impact craters also dot the surface.

Venus is the slowest-spinning planet, taking longer to spin than to complete an orbit of the Sun. Uniquely, it spins from east to west.

Cloudy world
Venus has a dense carbon-dioxide atmosphere and is completely shrouded by thick, highly reflective clouds made of dilute sulphuric acid droplets. These start about 45km (28 miles) above the surface and extend up to a height of 70km (43 miles). Below the clouds is an overcast, orange-coloured world.

Saskia Crater
Impact craters on Venus vary from 270km (168 miles) down to 7km (4 miles) across. Saskia is middle-sized, at 37km (23 miles), and has a central mountain. This view made by radar has been coloured based on images taken on the surface.

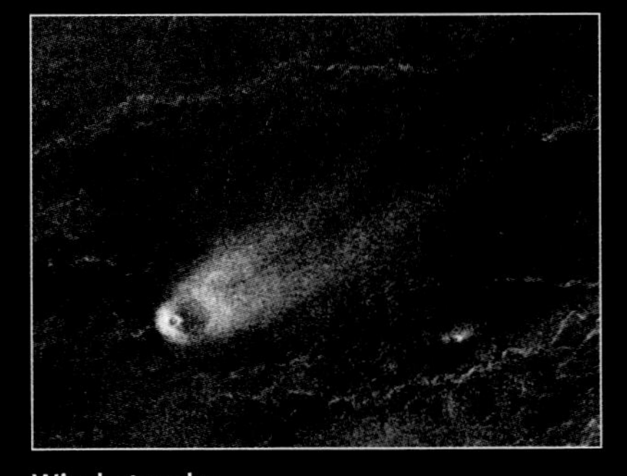

Wind streaks
The surface of Venus is windy, and since the winds generally blow in only one direction, wind streaks form. The streaks of dust and soil shown here are 35km (22 miles) long.

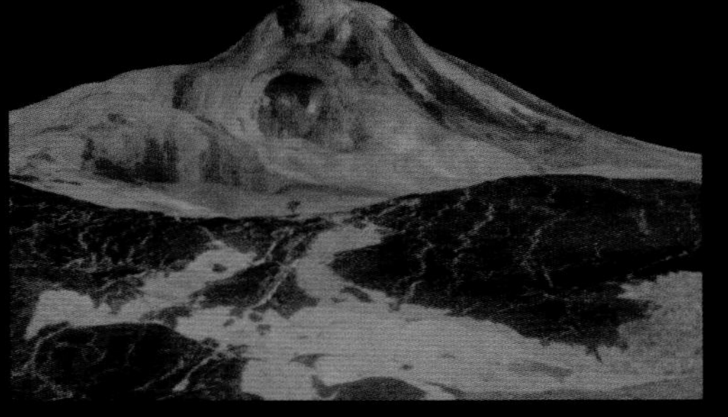

Maat Mons
This large shield volcano rises about 5km (3 miles) above the surrounding plains. Lava has poured out of the volcano, spread out in all directions for hundreds of kilometres, and then solidified.

PLANET DATA

DIAMETER	*12,104km (7,521 miles)*
AVERAGE DISTANCE FROM SUN	*108.2 million km (67.2 million miles)*
ORBITAL PERIOD	*224.7 Earth days*
ROTATION PERIOD	*243 Earth days*
SURFACE TEMPERATURE	*464°C (867°F)*
NUMBER OF MOONS	*none*

Binocular view
Venus is extremely bright when viewed through binoculars – its crescent phase is apparent – but all that is seen are the tops of the thick, cloudy blanket.

Phases of Venus
Because its distance from Earth changes with its various phases, Venus appears three times bigger when in crescent phase than it does when full.

Transit of Mercury
Mercury's orbit usually takes it above or below the Sun's disc in the sky. Infrequently, it is seen to transit the Sun's face. It can take up to 9 hours for Mercury to cross the face of the Sun. The black planetary dots on this image are Mercury's transit on 7 May 2003. The transit shows just how small Mercury is in comparison to the Sun. The next transit will take place on 9 May 2016.

PLANET DATA

DIAMETER *6,780km (4,213 miles)*

AVERAGE DISTANCE FROM SUN *227.9 million km (141.6 million miles)*

ORBITAL PERIOD *687 Earth days*

ROTATION PERIOD *24.6 hours*

SURFACE TEMPERATURE *–125 to +25°C (–193 to +77°F)*

NUMBER OF MOONS *2*

Giant features
A complex system of canyons, the Valles Marineris, slices Mars's surface and runs for more than 4,000km (2,500 miles). The three dark spots on the left are giant volcanoes on the Tharsis Bulge.

Mars

Mars is the next planet out from the Sun after Earth, and the outermost of the four rocky planets. It is a dry, cold world with a landscape marked by deep canyons and towering volcanoes. Mars is about half the size of Earth, and like Earth it has polar ice caps and seasons, and it spins around in a little over 24 hours.

FEATURES

Mars is made mainly of rock with a small, probably solid iron core. Its rocky surface has been moulded by faulting, volcanism, meteorites, water, and wind. Large-scale features such as the Valles Marineris were formed billions of years ago when internal forces split the surface. Elsewhere, huge regions such as the Tharsis Bulge were raised up above the surrounding terrain. The Tharsis Bulge is Mars's main volcanic centre and home to giant volcanoes, including Olympus Mons.

Low-lying lava plains cover much of the northern hemisphere. The highland south is older and heavily cratered from meteorite bombardment some 3.9 billion years ago. Dry riverbeds, outflow channels, and floodplains indicate that Mars once had flowing water.

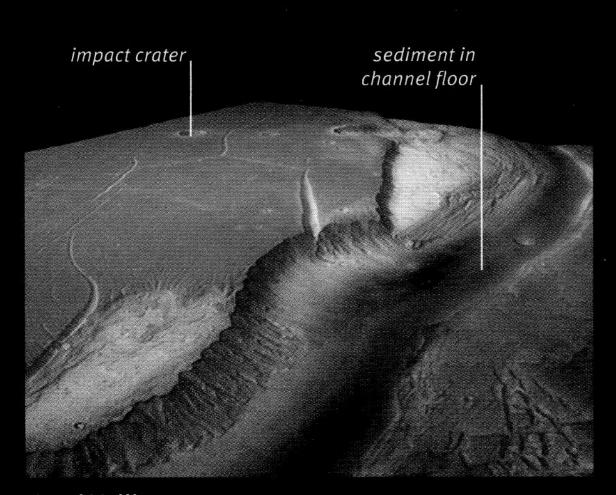

Kasei Vallis
Features such as Kasei Vallis bear witness to the presence of large amounts of fast-flowing water 3–4 billion years ago. This outflow channel a few hundred kilometres wide was probably formed by catastrophic flooding and glacial activity.

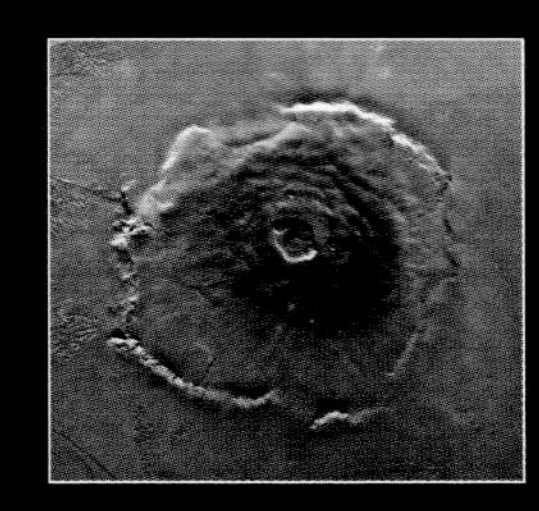

Olympus Mons
This is an overhead view of the summit of Olympus Mons. At 24km (15 miles) high, it is the biggest volcano in the Solar System.

Red planet
Much of Mars is rocky, sandy, and dusty, and large areas of its terrain resemble a rock-strewn desert. It is often referred to as the red planet, with its colouring coming from iron oxide (rust) in the rocks and soil.

OBSERVATION

Mars is in Earth's sky for much of each year and is one of the easiest planets to see. It is prominent and red and can be spotted with the naked eye by anyone with good eyesight.

LOCATING AND LOOKING

The best time to see Mars is when the planet is close to Earth and on the opposite side of Earth to the Sun (termed "opposition"; see p.26). This is when it is at its brightest and largest; it is also above the horizon all night long. Oppositions occur roughly every other year, since they are about 26 months apart. Dates of oppositions are included in the Special Events listings in the Monthly Sky Guide (pp.96–121). All oppositions are good for observing Mars, but some are better than others. This is because Mars follows an elliptical orbit around the Sun, and so its distance from us varies. Mars's brightness at opposition is in the range –1.0 to –2.8.

Mars keeps close to the path of the ecliptic and is found within the zodiac band of sky. It is only some 78 million km (48.5 million miles) away from Earth at opposition and makes rapid progress against the background stars. It travels west to east but goes into retrograde motion (see p.26) about every 22 months. At such a time, Mars appears to move backwards for a few weeks as Earth passes between it and the Sun before once again resuming its forward progression. Mars starts its backward motion about five weeks before opposition.

Mars is the only rocky planet with surface features visible from Earth. Its polar caps can be seen as, in turn, they are tilted towards Earth. Light and dark areas can be seen on the disc; these do not relate to real surface features but result from differences in reflectivity (dark areas reflect the least).

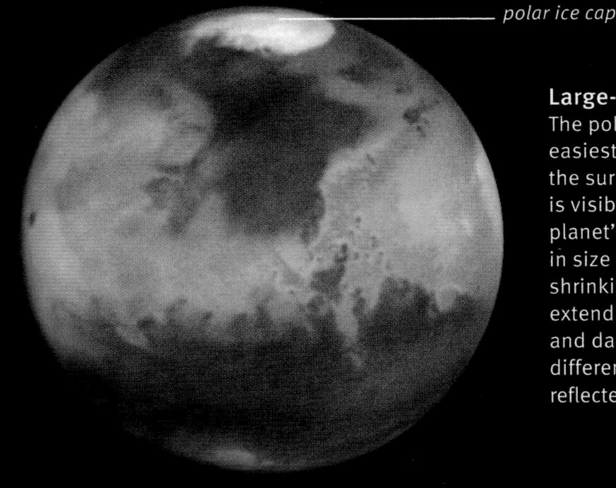

Large-telescope view
The polar ice caps are the easiest features to pick out on the surface of Mars. Just one is visible at a time due to the planet's tilt. The caps change in size with the seasons, shrinking in summer and extending in winter. Light and dark marks (the result of differences in the amount of reflected sunlight) can be seen.

Naked-eye view
Mars is identified by its disc shape and red colouring. Here, it is the brighter of the two dots. Jupiter is to its right.

Binocular view
The planet's disc shape becomes obvious through binoculars, but its surface features are still not visible.

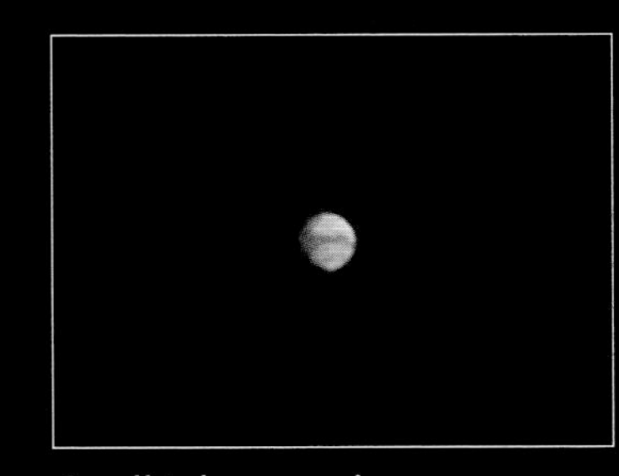

Small-telescope view
Here, Mars is visibly red. Some surface features – polar ice caps and dark regions – may be seen.

PLANET DATA

DIAMETER *142,984km (88,846 miles)*

AVERAGE DISTANCE FROM SUN *778.3 million km (483.6 million miles)*

ORBITAL PERIOD *11.8 Earth years*

ROTATION PERIOD *9.9 hours*

CLOUD-TOP TEMPERATURE *−110°C (−166°F)*

NUMBER OF MOONS *63*

Storm world
Jupiter's cloudy atmosphere is dominated by storms. The smallest are like Earth's largest hurricanes. The biggest, the Great Red Spot, is bigger than Earth itself. On the left is the shadow of the moon Europa.

Jupiter

Jupiter is a giant among the planets. It is the second-largest body in the Solar System after the Sun and the most massive of all the planets. Its visible surface is not solid but the colourful top layer of a deep, thick atmosphere. An extensive family of moons orbits around Jupiter, and a thin faint ring of dust particles encircles it.

FEATURES

Jupiter is made of hydrogen and, to a lesser extent, helium. Its outer layer is a 1,000km- (600-mile-) deep hydrogen-rich atmosphere. Underneath this, the planet becomes increasingly dense and hot, and the hydrogen acts like a liquid; deeper still, it acts like a molten metal. At Jupiter's heart is a core of rock, metal, and hydrogen compounds.

The "surface" we see consists of colourful bands of swirling gas. Heat from inside Jupiter combined with its fast spin create the bands and the violent weather within them. The light bands of rising gas and the red-brown ones of falling gas produce winds that give rise to storms and hurricanes. The Great Red Spot is a giant storm that has raged for more than 300 years.

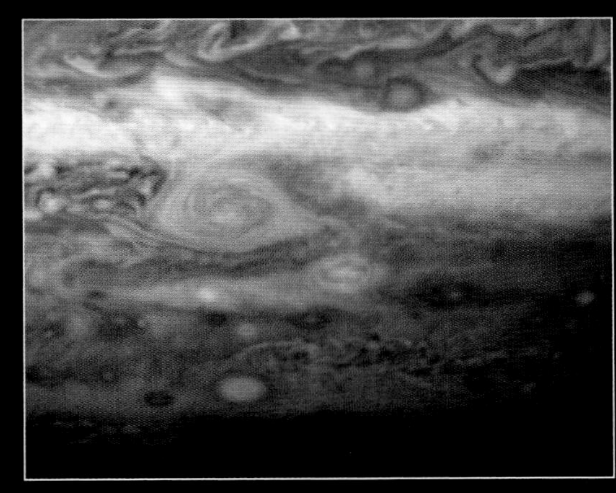

Little Red Spot
In late 2005, astronomers witnessed the birth of Jupiter's second red spot, and second-largest storm – about 70 per cent of the size of Earth. It formed between 1998 and 2000 when three white, oval-shaped storms merged.

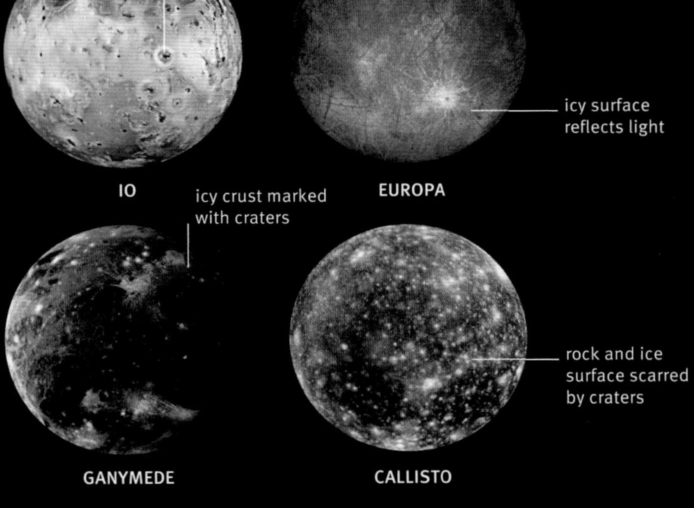

Galilean moons
Jupiter's four largest moons are collectively known as the Galilean moons. These spherical bodies made of a mix of rock and ice are worlds in their own right. By contrast, most of the other moons are small and irregular in shape.

OBSERVATION

Jupiter shines like a bright silver star in the night sky for about ten months of each year. Its brilliance allows it to stand out from the star background and be recognized. Although its brightness varies, it is never dimmer than Sirius, the brightest star in the night sky.

LOCATING AND LOOKING

Jupiter is best seen when it is close to Earth and on the opposite side of Earth to the Sun (a position known as "opposition"; see p.26). It is then particularly bright because the sunlight that shines fully on it is reflected back by its atmosphere. Its brightness at opposition is at least –2.3 and goes up to a maximum of –2.9. Jupiter is also in the sky all night long, rising at sunset, being highest in the middle of the night, and setting at sunrise. Opposition occurs every 13 months; dates of oppositions are included in the Special Events listings in the Monthly Sky Guide (pp.96–121).

Jupiter is found within the zodiac band of sky. It takes about 12 months to cross one zodiac constellation before moving into the next. It travels from west to east against the background of stars, but when at opposition it also goes through a period of retrograde motion (see p.26) and temporarily moves backwards in the sky.

For most of the time, Jupiter is the planet with the largest disc size. Surface detail on the planet can be seen through either powerful binoculars or a telescope. The banded structure, the Great Red Spot, and other cloud features then come into view. A large telescope will reveal further spots and structure in the clouds. The slightly squashed appearance of Jupiter's disc, due to its fast spin, is also apparent through a telescope.

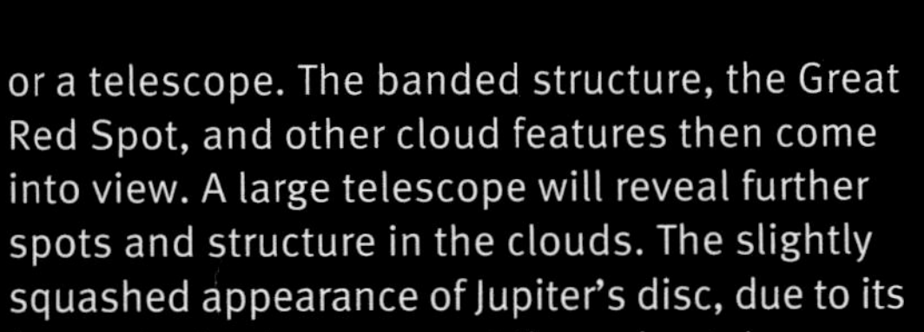

Changing view
Our view of Jupiter's surface is constantly changing due to the planet's rapid rotation. Jupiter makes one spin in just under ten hours. This means that features can be seen to move within about ten minutes. These five images show Jupiter over the course of about five hours. The planet spins from left to right, and the Great Red Spot is seen to move as the planet turns. The black dot is the shadow of a Galilean moon.

Naked-eye view
Jupiter's disc shape is clearly visible even to the naked eye. Its shape and brilliance make it easy to identify.

Binocular view
The Galilean moons are visible through binoculars either side of the planet's equator, changing position each night.

Telescope view
Here, it is possible to discern details: the bands and the Great Red Spot can be seen.

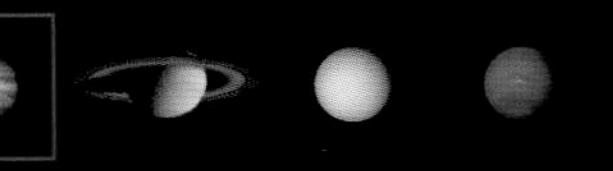

Saturn

Pale-yellow Saturn is the most distinctive planet due to its impressive ring system. It is second in size to Jupiter. Like Jupiter, its visible surface is its outer atmosphere, and it has a large family of moons. Saturn is also the most distant planet that is normally visible to the naked eye.

Saturn and its moons
Saturn dwarfs two of its largest moons, Tethys (top) and Dione (below). Shadows cast by Tethys and the main rings are seen on the planet's globe. The limb of the planet is also visible through the Cassini Division separating the A and B rings.

FEATURES

Saturn is made mainly of hydrogen and helium. They form the planet's gaseous outer layer, but inside, as the temperature and pressure increase with depth, the hydrogen and helium change state. Below the atmosphere they act like a liquid and, deeper still, like a liquid metal. A core of rock and ice is at the planet's centre.

Ringed world
Saturn's rings reflect sunlight well, making them and the planet easy to see. But those rings we readily see (shown here) are only part of the system; much fainter rings extend to about four times as far from the planet. Saturn is tilted relative to its orbit, so as it travels, first one hemisphere then the other is tilted sunwards. This gives Saturn seasons and offers a changing view from Earth.

PLANET DATA

DIAMETER *120,536km (74,897 miles)*

AVERAGE DISTANCE FROM SUN *1.43 billion km (886.56 million miles)*

ORBITAL PERIOD *29.5 Earth years*

ROTATION PERIOD *10.7 hours*

CLOUD-TOP TEMPERATURE *–180°C (–292°F)*

NUMBER OF MOONS *56*

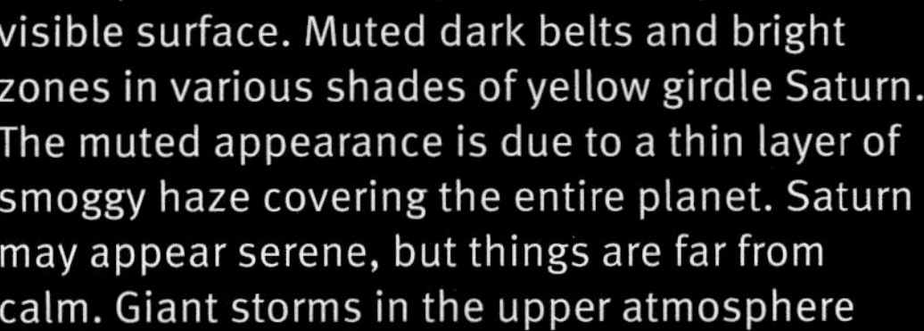

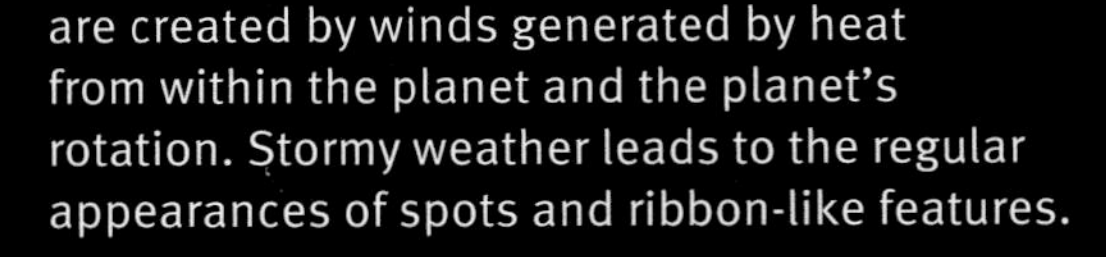

The top of the cloud layer forms the planet's visible surface. Muted dark belts and bright zones in various shades of yellow girdle Saturn. The muted appearance is due to a thin layer of smoggy haze covering the entire planet. Saturn may appear serene, but things are far from calm. Giant storms in the upper atmosphere are created by winds generated by heat from within the planet and the planet's rotation. Stormy weather leads to the regular appearances of spots and ribbon-like features.

MOONS AND RINGS

Most of Saturn's moons are small and irregularly shaped. Only Titan is larger than our Moon. Most were discovered in the past 25 years, and more are expected. The moons are mixes of rock and water-ice in varying proportions. Some moons orbit within the ring system, which is not solid but is made of billions of pieces of dirty water-ice. These range in size from dust grains to objects several metres across, and they move around Saturn on their own orbits. The system extends for hundreds of thousands of kilometres into space, but it is only a few kilometres deep.

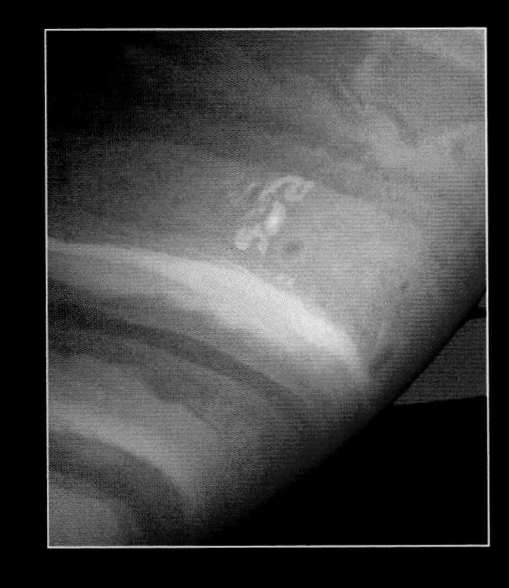

Dragon storm
In September 2004, an area in the southern hemisphere known for its storms and dubbed "storm alley" was the setting for a giant thunderstorm. Dragon Storm – the reddish feature above the centre in this false-colour image – seems to be a long-lived storm that periodically flares up.

C ring *B ring* *Cassini Division* *A ring*

Main rings
Saturn's readily seen ring system comprises three distinct rings: A, B, and C. These are made of ringlets, which in turn are made of pieces of dirty water-ice. The Cassini Division, seen from Earth as an empty gap, is in fact full of ringlets.

OBSERVATION

Saturn is twice as remote from Earth as Jupiter but is still bright enough to be seen at some time during the night about ten months of the year. It looks like a star but can be distinguished from its stellar backdrop with the help of the appropriate constellation card.

LOCATING AND LOOKING

Due to Saturn's distance from Earth, it moves more slowly across the sky than closer planets such as Jupiter. Its 29.5-year orbit means it takes about two and a half years to pass through one zodiac constellation. It travels from west to east, but every 12 months it goes into retrograde motion (see p.26), when, for about four months, it appears to move backwards.

Like other superior planets, Saturn is best seen when it is on the opposite side of Earth to the Sun (termed "opposition"; see p.26). Its brightness at opposition ranges from 0.8 to a maximum of −0.3. This wide range is largely due to the varying amount of ring facing Earth. Oppositions happen annually – about two weeks later each year. Dates of oppositions are included in the Special Events listings in the Monthly Sky Guide (pp.96–121).

A small telescope or powerful binoculars will reveal the nature of the rings and give a first look at the disc's banded appearance. A larger telescope reveals more disc detail, the three main rings, the Cassini Division, and a handful of dots, which are moons. The bulging equatorial region and flattened poles are also apparent. Our view of the rings changes as Saturn orbits the Sun. Their orientation changes from edge-on (when they are virtually invisible), to open (fully visible), and then edge-on again. The rings were open last in 2002. They will be fully open and seen from above in 2017.

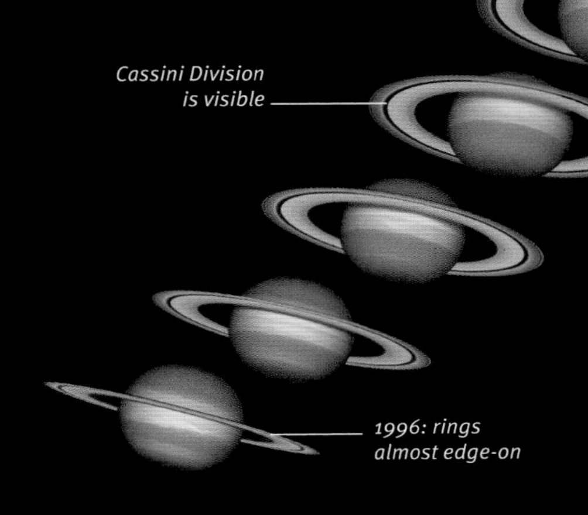

Changing view
Our view of the rings changes as Saturn and Earth move along their orbits. A full cycle is completed every 29.5 years. We see the rings edge-on first; from above as Saturn's north pole tilts towards the Sun; edge-on once again; and from below when the south pole is sunwards.

Naked-eye view
Saturn looks like a star to the naked eye, but its warm colouring and disc shape distinguish it from the stars.

Binocular view
Saturn's disc shape is more apparent here. High-powered binoculars will show the rings as bumps on its sides.

Small telescope view
The ring system can be clearly seen, and the planet's bands are just visible.

Uranus and Neptune

The two planets farthest from Earth, Uranus and Neptune are also the smallest of the gas giants. They are cold, featureless worlds, so remote that they are difficult to see from Earth. Both have sparse rings and large numbers of moons orbiting around them. And both are coloured blue by the presence of methane-ice clouds in their atmospheres.

URANUS

The visible surface of Uranus is its hydrogen-rich atmosphere. Below is a deep layer of water, methane, and ammonia ices, and in the centre a core of rock and possibly ice. Uranus appears calm and featureless because of upper-atmosphere haze, but the planet does undergo change. The atmosphere has some banding, and bright clouds are carried around the planet; in 2006, a dark spot was observed. Uranus's unique feature is its tilt: the planet appears to orbit the Sun on its side. The combination of the tilt and its 84-year orbit mean that its hemispheres face the Sun for 42 years at a time.

PLANET DATA

DIAMETER *51,118km (31,763 miles)*

AVERAGE DISTANCE FROM SUN *2.87 billion km (1.78 billion miles)*

ORBITAL PERIOD *84 Earth years*

ROTATION PERIOD *17.2 hours*

CLOUD-TOP TEMPERATURE *–214°C (–353°F)*

NUMBER OF MOONS *27*

Ring system
Uranus has two distinct sets of rings. Closest to the planet is a set of 11 rings; narrow and widely separated, these are more gap than ring. Outside these is a pair of faint, dusty rings.

NEPTUNE

Neptune is similar in structure to Uranus. It has a core of rock and possibly ice, surrounded by a mix of water, methane, and ammonia ices, and topped off with a hydrogen-rich atmosphere. The planet undergoes seasonal change, and its atmosphere is unexpectedly dynamic. It experiences ferocious equatorial winds, fast-moving bright clouds, and short-lived gigantic storms.

The planet is surrounded by a sparse system of five complete rings and one partial ring. Only one of its 13 moons, Triton, is of notable size, and four are within the planet's ring system.

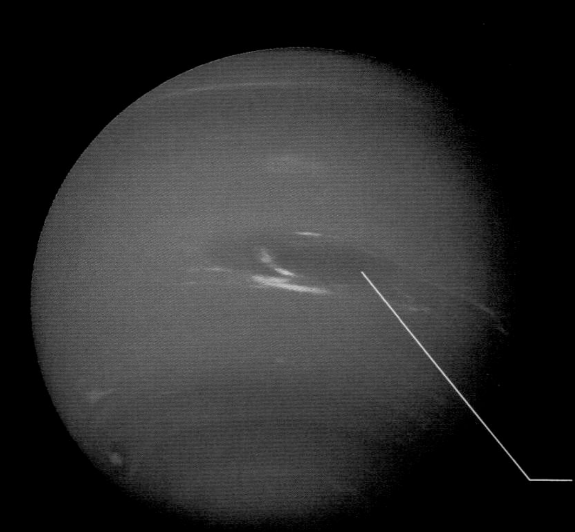

PLANET DATA

DIAMETER *49,532km (30,760 miles)*

AVERAGE DISTANCE FROM SUN *4.5 billion km (2.8 billion miles)*

ORBITAL PERIOD *164.9 Earth years*

ROTATION PERIOD *16.1 hours*

CLOUD-TOP TEMPERATURE *–200°C (–320°F)*

NUMBER OF MOONS *13*

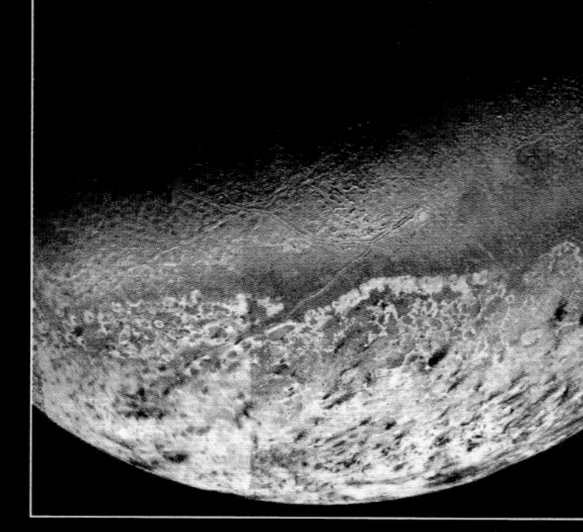

Triton
Triton is about three-quarters the size of Earth's Moon. It is a rocky world with an icy surface of linear grooves, ridges, and circular depressions.

OBSERVING THE TWO PLANETS

Both Uranus and Neptune are too distant to be seen easily. The sharp-sighted can see Uranus with the naked eye, but it is difficult to locate. Following its progress against the stars will confirm its identity. This will take time and patience, because Uranus's orbit is long, and the planet spends about seven years in the same zodiac constellation. Neptune moves even more slowly through the zodiac constellations, spending nearly 14 years in each. It is too faint to be seen by the naked eye, but a medium-sized telescope will reveal a blue-green disc.

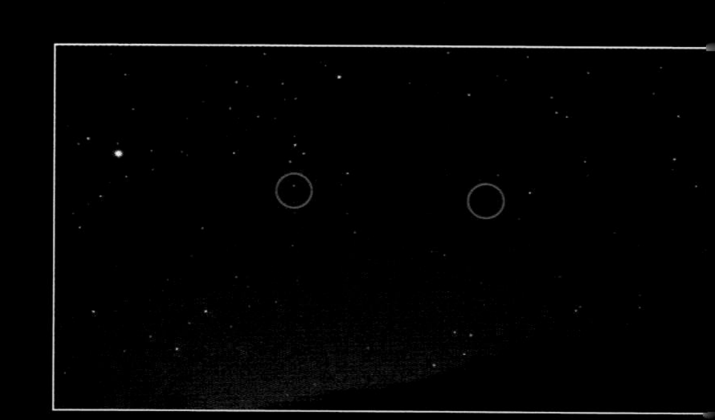

Uranus and Neptune
Distant Uranus (left) is magnitude 5.5 and brighter than Neptune (right), which, at magnitude 7.8, is beyond naked-eye vision. Both appear as faint stars in this enhanced view.

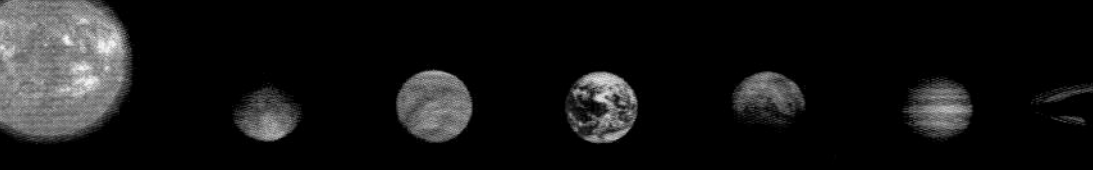
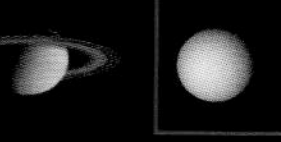

The dwarf planets

ris, Pluto, and Ceres are the three dwarf planets in the Solar System. They have existed or 4.6 billion years but have only been categorized as dwarf planets since 2006. The distant, cy worlds of Eris and Pluto orbit beyond Neptune. Ceres is much closer to the Sun in the lain Belt of asteroids.

WARF WORLDS

is is the largest dwarf planet, and it is believed be a mix of rock and ice with an icy surface. Its ze is uncertain, with estimates ranging from 400km (1,500 miles) to around 3,000km (1,850 iles) across. Eris was discovered in 2005 when nages taken in 2003 were reanalyzed. At the me, it was almost 16 billion km (10 billion miles) way and the most distant Solar System object een. It completes one circuit of its elongated nd highly inclined orbit every 560 years.

Pluto is a cold world. A thin, icy crust covers a ody made of probably 70 per cent rock and 30 er cent water-ice. At an average distance from e Sun of 5.9 billion km (3.7 billion miles), it is nsurprising that its surface temperature is bout –230°C (–382°F). This 2,304km- (1,432- ile-) wide body has three moons. The largest, haron, is about half the size of Pluto itself. ach of the pair spins every 6.38 days, and ecause Charon orbits Pluto in the same time, e two bodies keep the same face towards each ther. From its discovery in 1930 until August 006, Pluto had planetary status. It was then e most distant and smallest of the planets.

Ceres has been known since 1801. It was the rst asteroid discovered and remains the largest nown; it measures 960km (596 miles) across. eres is a rocky body with water-ice near the urface. It orbits the Sun every 4.6 years.

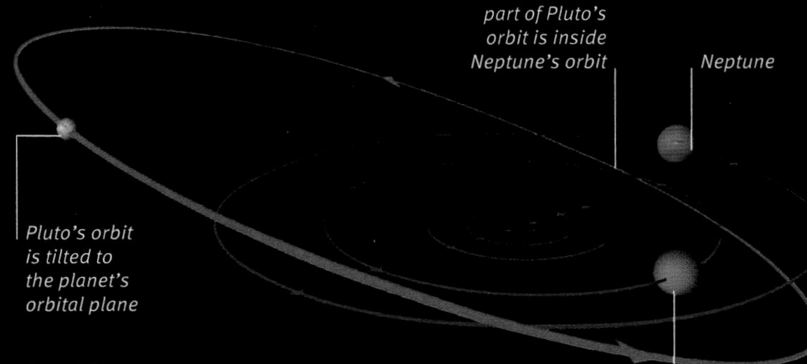

Pluto's orbit
The orbit of Pluto is elongated and tilted with respect to the orbits of the eight planets. Pluto completes one circuit in about 248 years, and for about 20 years of each orbit its path takes it closer to the Sun than Neptune. This last happened between 1979 and 1999.

Ground view
This is how Pluto appears through the best ground-based telescopes. The bump to the lower left is its major moon, Charon.

PLANETS OR DWARF PLANETS?

After the first Kuiper Belt object was discovered, in 1992, Pluto's planetary status was questioned. The 2005 identification of an object larger than Pluto led to the creation of the "dwarf planet" class. This was established in August 2006 by the International Astronomical Union, the largest professional body for astronomers. Like planets, dwarf planets are almost round, but unlike planets, they have not cleared their neighbourhoods: Ceres is in the Main Belt; Eris and Pluto are in the Kuiper Belt. So far, only these three examples are known, but other likely candidates exist in the Kuiper Belt, and the number will rise.

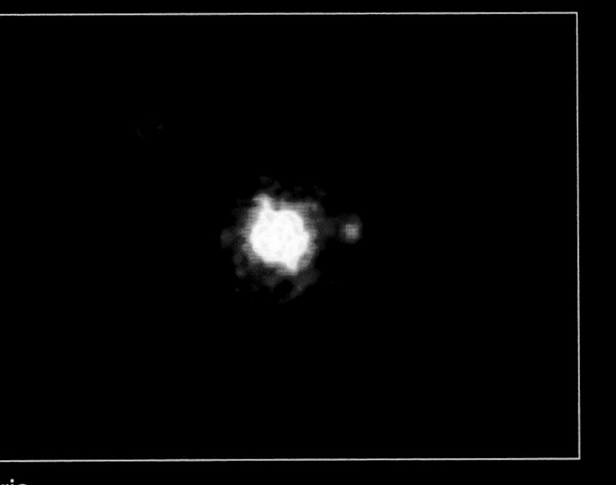

ris
is image from the 10m (3ft) Keck telescope in Hawaii has Eris t its centre as a large bright object. The smaller, duller object its right is Dysnomia. Keck astronomers realized the small ot was a moon when it moved with Eris against the stars.

Ceres
The Hubble Space Telescope studied Ceres for nine hours, the time it takes to make one orbit. It was found to be almost round: the diameter at its equator is wider than at its poles.

Pluto and its moons
On average, Pluto is a little under 40 times as far from the Sun as Earth. This image from the Hubble Space Telescope shows Pluto with Charon at its lower right and, further right, the moons Nix (top) and Hydra (bottom).

Comet McNaught
On 7 August 2006, Robert McNaught discovered a comet that within months became the brightest in Earth's sky for more than 30 years. In January 2007, when at its closest to the Sun, it had a large head and spectacular tail and was impossible to miss in the southern sky.

Comet Tempel 1
This image was taken in 2005 by the *Deep Impact* spacecraft 67 seconds after the craft released a missile to impact the nucleus. Light from the collision shows a pitted, ridged surface.

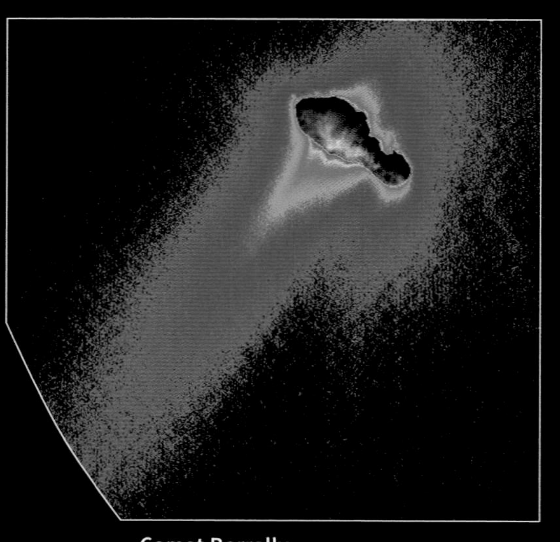

Comet Borrelly
The bowling-pin-shaped nucleus of Comet Borrelly was imaged by the *Deep Space 1* spacecraft in 2001. It is 8km (5 miles) long and orbits the Sun every 6.9 years. Colour has been added to highlight dust and gas jetting out of the nucleus.

Comets, meteors, and asteroids

Billions of comets and asteroids are invisible to us on Earth because they are too distant or too small to be seen. Yet some comets appear to grow and then make a spectacular show in the night sky. On any given night, meteors produced by cometary dust flash into view, and asteroids can crash-land on Earth's surface.

COMETS

Comets are often referred to as dirty snowballs. But these cosmic snowballs are not ball-shaped, and they are on a giant scale. Comets are irregular-shaped, city-sized lumps of snow and rocky dust that follow orbits around the Sun. More than a trillion of them together make the vast Oort Cloud, where they have been since the birth of the Solar System 4.6 billion years ago.

Comets only become visible when they leave the Oort Cloud and travel in towards the Sun. Then, the snowball, called a nucleus, is warmed by the Sun's heat. The snow is converted into gas, and this and loosened dust jet away from the nucleus. When closer to the Sun than Mars, the nucleus is surrounded by a cloud of gas and dust, known as a coma, and it develops two tails.

The comet is now large enough and bright enough to be seen from Earth. Typically, two or three a year are seen as a fuzzy patch of light through binoculars. Three or four a century, such as Comet McNaught, are truly impressive and easily visible to the naked eye.

More than 2,000 comets have been recorded within the vicinity of the Sun, and about 200 are periodic. Most are named after their discoverers.

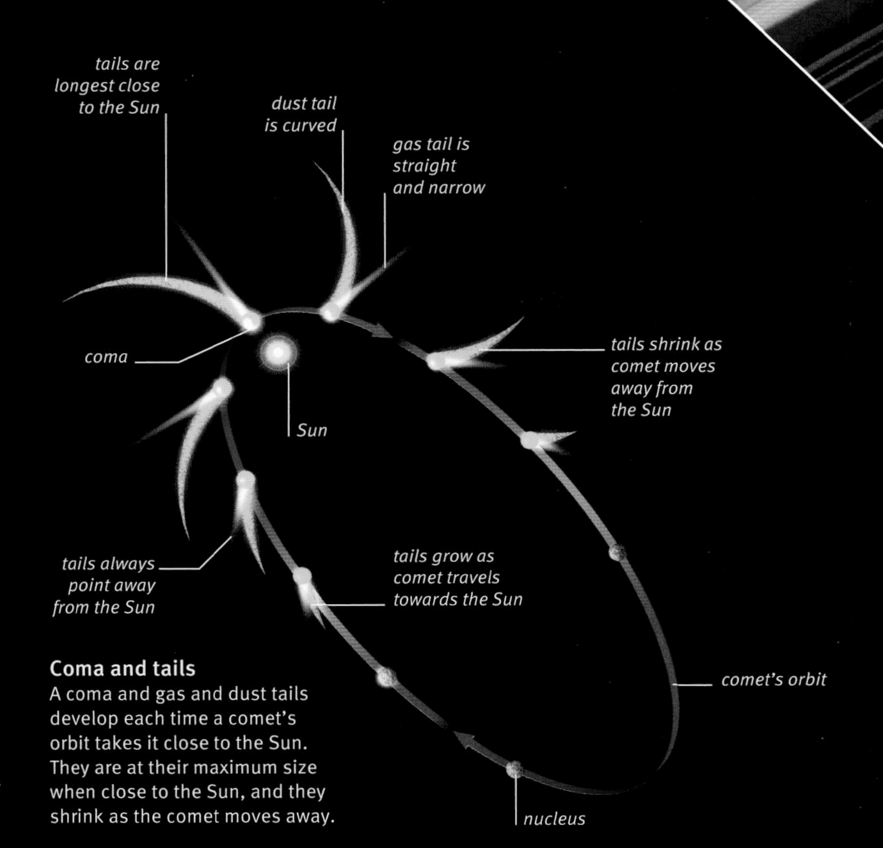

Coma and tails
A coma and gas and dust tails develop each time a comet's orbit takes it close to the Sun. They are at their maximum size when close to the Sun, and they shrink as the comet moves away.

METEORS

On any night of the year it is possible to see a meteor – a short-lived trail of light in the sky. These transient flashes, popularly called shooting stars, are produced by small fragments of a comet or, sometimes, an asteroid. As the fragment, or meteoroid, moves through Earth's atmosphere, it produces a trail of excited atoms, which in turn produce light. This streak of light is the meteor. It lasts for less than a second and has an average brightness of magnitude 2.5.

The best time to see a meteor is during the hours before dawn, when you'll be on the side of Earth facing the incoming meteoroids. Although meteors appear every night, they are not predictable. To maximize your chances of seeing one, it is best to observe when a meteor shower is expected. These happen on the same dates every year and are produced as Earth ploughs into a stream of comet particles. Shower details are in the Monthly Sky Guide (see pp.96–121).

Leonid shower
The Leonid meteor shower occurs in mid-November. As Earth orbits, it passes through a stream of particles lost by Comet Tempel–Tuttle. The dust particles produce meteors that appear to radiate from a point in the constellation of Leo.

ASTEROIDS

Asteroids are dry, dusty lumps of rock that failed to make a rocky planet. Most are irregular in shape, and just the handful or so over about 320km (200 miles) across are round. They range in size from Ceres at 960km (596 miles), through boulders and pebbles to dust-sized particles. There are 100,000 larger than about 20km (12.5 miles) across. As size diminishes, the quantity increases. Most orbit the Sun between Mars and Jupiter.

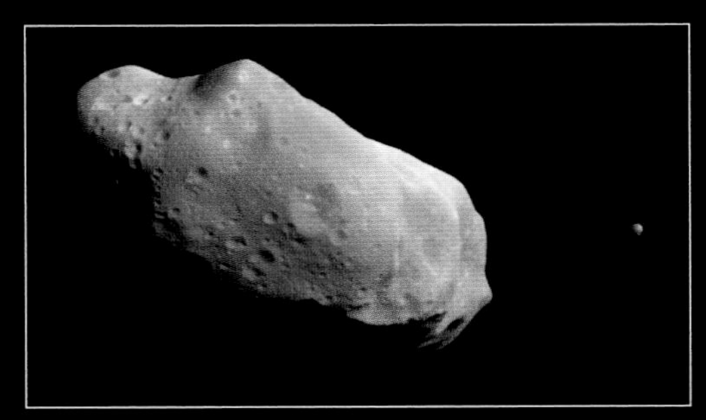

Ida and Dactyl
The asteroid Ida is 60km (37 miles) long and shows signs of past collisions. It journeys around the Sun every 4.8 years. Its tiny moon, Dactyl, just 1.6km (1 mile) long, orbits Ida every 27 hours.

Meteorites
When an asteroid makes it through Earth's atmosphere and lands on the surface, it is called a meteorite. More than 22,500 have been catalogued. This one has a dark crust that formed as it fell through the atmosphere.

AQUARIUS
PISCES
Andromeda
Pegasus
Equuleus
Delphinus
CAPRICORNUS
Cassiopea
Cepheus
Cygnus
Antinous
Aquila
Lira
SAGITTARIUS
Vrsa minor
Engonasi
Hercules
Draco
Corona Bor.
Serpentarius
Ophiuchus
Bootes
Serpens
Arcturus
Coma Berenices
SCORPIUS
LIBRA
VIRGO
HYPOTHESIS PTOLEMAICA
HYPOTHESIS COPERNICANA
CAPRICORNUS
Grus
Indus
Pauo
Corona australis
SAGITTARIUS
SCORPIUS
Ara, Thuribulum
Fera, Lupus
LIBRA
Sol

Mapping the sky

Astronomers divide the sky into 88 areas called constellations, which are interlocked like pieces of an immense jigsaw puzzle. Each of the 88 constellations are covered in this section, with a chart and a text profile that includes a list of stars and other features that can be found within its borders. Guidance on when to view them is given in the Monthly Sky Guide chapter (see pp.96–121).

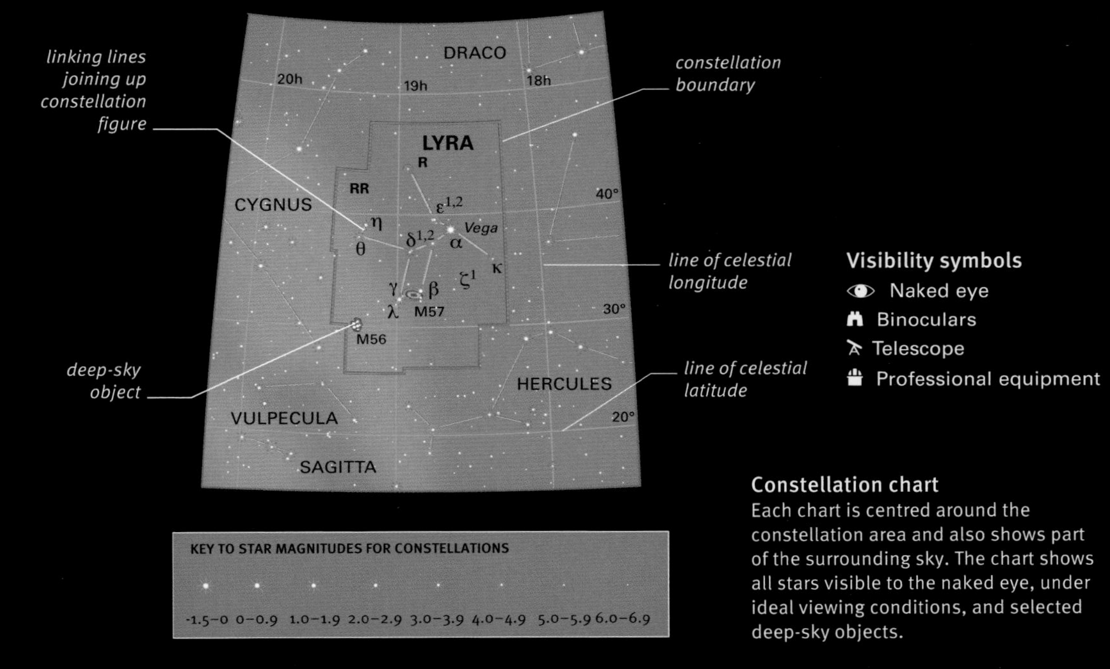

Visibility symbols

- Naked eye
- Binoculars
- Telescope
- Professional equipment

Constellation chart
Each chart is centred around the constellation area and also shows part of the surrounding sky. The chart shows all stars visible to the naked eye, under ideal viewing conditions, and selected deep-sky objects.

Fully Visible
Partially visible
Not visible
80°N
60°N
40°N
20°N
0°
20°S
40°S
60°S

Visibility map
The chart for each constellation is accompanied by a map that shows the parts of the world from which it can be seen. The entire constellation can be seen from the area shaded in green, and part of it can be seen from the area shaded in yellow, but it cannot be seen at all from the area shaded in red.

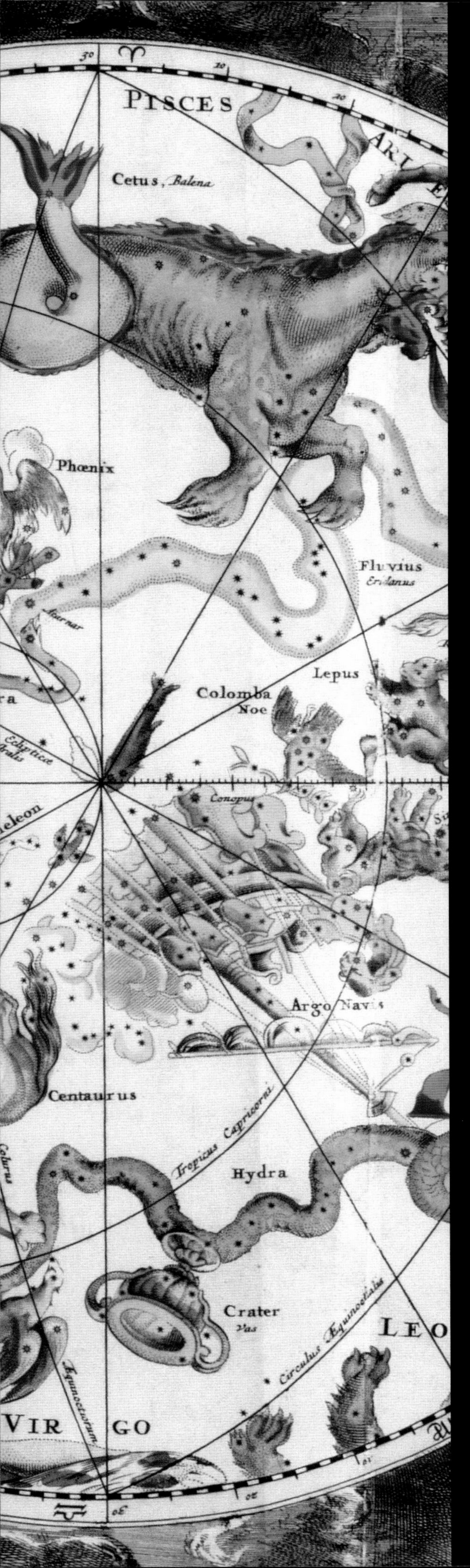

A view of the heavens
Since ancient times, the stars in the sky were associated with gods, legends, heroes, and mythical beasts. This 15th-century chart shows the way these figures were represented in the northern and southern halves of the night sky.

Long-tailed bear
The tail of the Little Bear curves away from the north Pole Star, Polaris (top left). Unlike real bears, the celestial bears Ursa Minor and Ursa Major both have long tails.

Ursa Minor The Little Bear

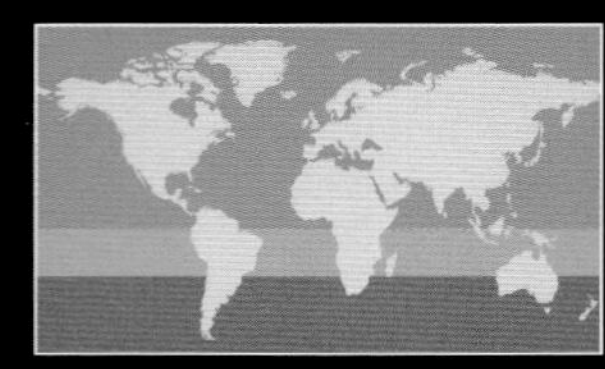

Fully visible 90°N–0°

The constellation of Ursa Minor is a constant feature of northern skies – as the location of the north celestial pole, it never rises or sets, but instead spins around the pole once every 24 hours. Its shape mimics that of the Plough (or Big Dipper) in Ursa Major.

FEATURES OF INTEREST

Alpha (α) Ursae Minoris (Polaris) The northern pole star, Polaris is an almost fixed point in the sky, since it lies barely half a degree from the north celestial pole. A yellow supergiant star of magnitude 2.0, Polaris lies about 430 light years away. Until quite recently, it was classed as a Cepheid variable – a type of pulsating star named after its prototype, Delta Cephei in neighbouring Cepheus. But in the past few decades its variations have died away – a sign that occasionally changes in stellar evolution can happen within a human lifetime. Polaris's 8th-magnitude companion star can be picked up with a small telescope.

Beta (β) Ursae Minoris (Kochab) Ursa Minor's second brightest star is an orange giant about 100 light years away.

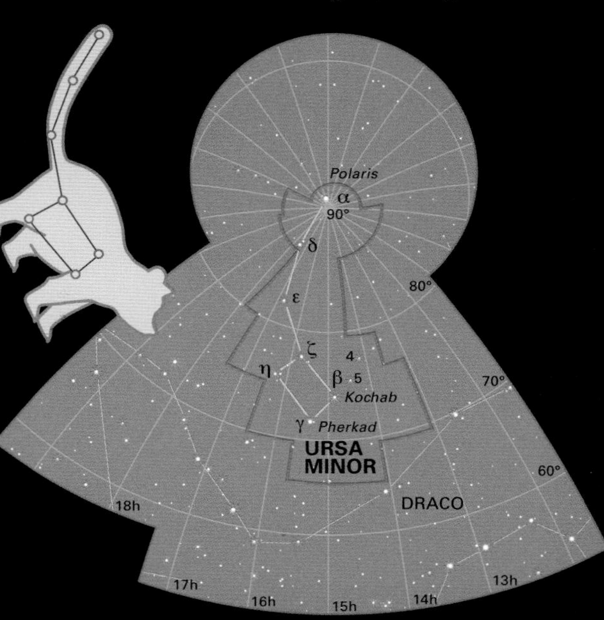

Draco The Dragon

Fully visible 90°N–4°S

The large constellation of Draco wraps itself around Ursa Minor and the north celestial pole. It represents a dragon from Greek mythology, slain by Hercules. Despite its size, it has no stars brighter than second magnitude.

FEATURES OF INTEREST

Alpha (α) Draconis (Thuban) Thuban is a blue-white giant star, about 300 light years from Earth. Precession, the slow wobble of Earth's axis of rotation, meant that 5,000 years ago, this star was the pole star.

16 and 17 Draconis This pair of stars, magnitudes 5.1 and 5.5, are easily divided with binoculars, but a small telescope shows that the brighter star is itself a double.

NGC 6543 This is one of the sky's brightest planetary nebulae.

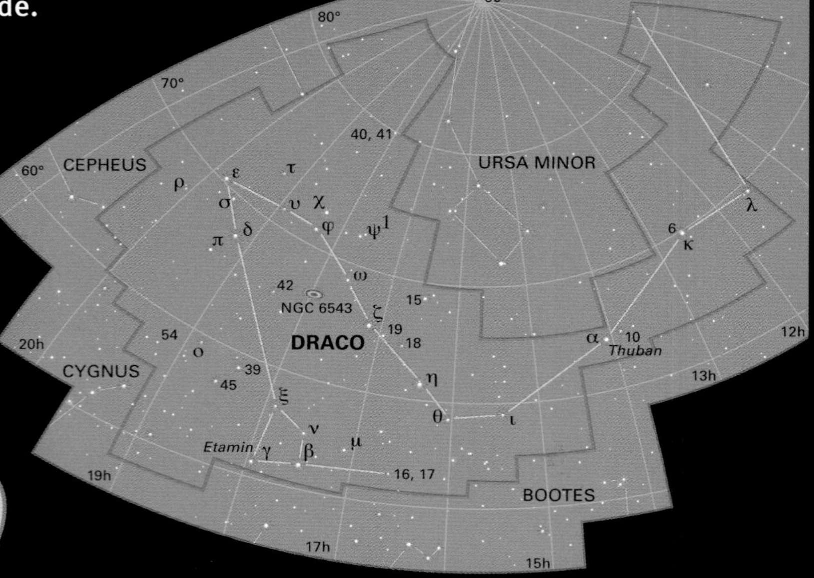

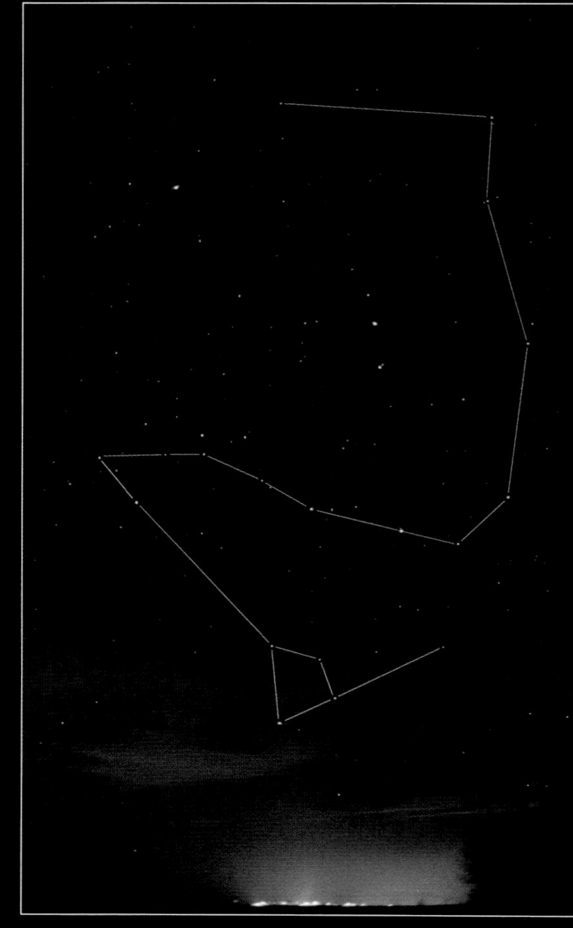

Dragon star
The lozenge-shaped head of the dragon is easily identified in the constellation Draco. This shape is formed by four stars, including Gamma (γ) Draconis, the brightest star.

Cepheus Cepheus

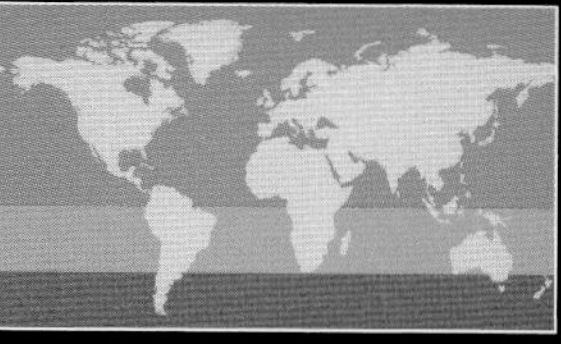

Fully visible 90°N–1°S

Lying in the far northern sky between Draco and Cassiopeia, this constellation represents the husband of Queen Cassiopeia in Greek myth. Although its pattern is obscure, it contains several interesting variable stars.

FEATURES OF INTEREST

Beta (β) Cephei The constellation's second brightest star is a blue giant with a faint companion. Its brightness varies in a 4.6-hour cycle, but only by 0.1 magnitudes.

Delta (δ) Cephei The prototype star for a class of variables called Cepheids, this ageing yellow supergiant is passing through a phase of its life where it expands and contracts repeatedly. It changes brightness between magnitudes 3.5 and 4.4 in a little under five days and nine hours.

Mu (μ) Cephei Called the Garnet Star due to its blood-red colour, Mu is a red supergiant. Like Delta, it is a variable star, but it is less predictable, varying between magnitudes 3.4 and 5.1 with a period of about two years.

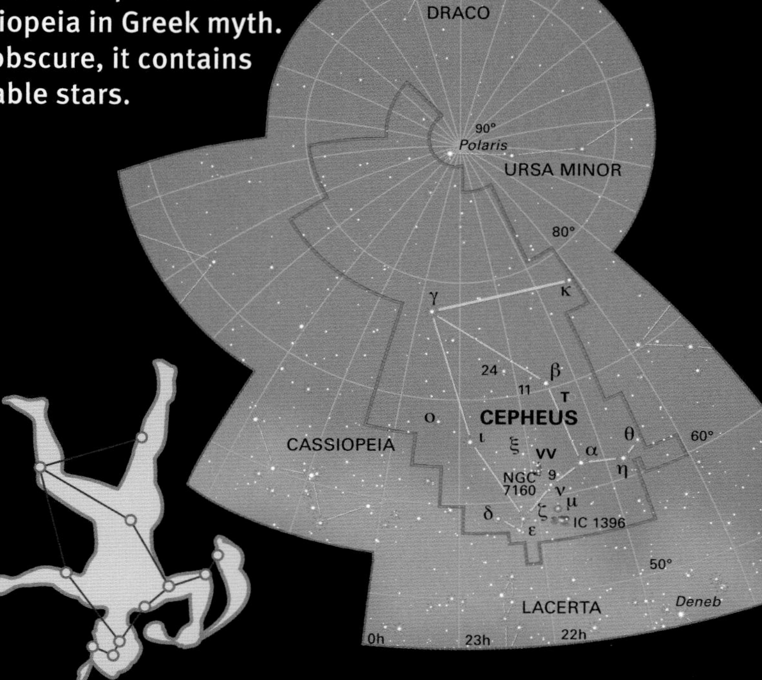

The king
Shaped like a bishop's mitre, Cepheus is not easy to pick out in the sky. He is flanked by his prominent wife, Cassiopeia, and Draco, the dragon.

Cassiopeia Cassiopeia

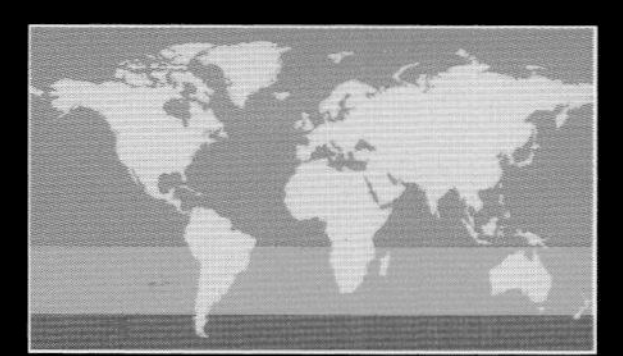

Fully visible 90°N–12°S

This distinctive W-shaped group of stars sits on the opposite side of the pole star from the Plough or Big Dipper, and is usually easy to locate. One of the original Greek constellations, it depicts Cassiopeia, the mother of Andromeda in Greek myth, sitting in a chair and fussing with her hair.

FEATURES OF INTEREST

Alpha (α) Cassiopeiae (Shedir) The brightest star is a yellow giant of magnitude 2.2, 120 light years away.

Gamma (γ) Cassiopeiae Gamma is one of the youngest stars obvious to the naked eye. It lies about 800 light years away, and usually shines at magnitude 2.5. However, it is still expelling material from the nebula in which it formed, and the obscuring gas can cause its brightness to vary unpredictably between magnitudes 3.0 and 1.6.

NGC 457 Cassiopeia's position in the heart of the northern Milky Way means that it is rich in star clusters, the best of which is NGC 457. This ball of 80 recently formed stars, 9,000 light years from Earth, is visible to the naked eye and rewarding in binoculars.

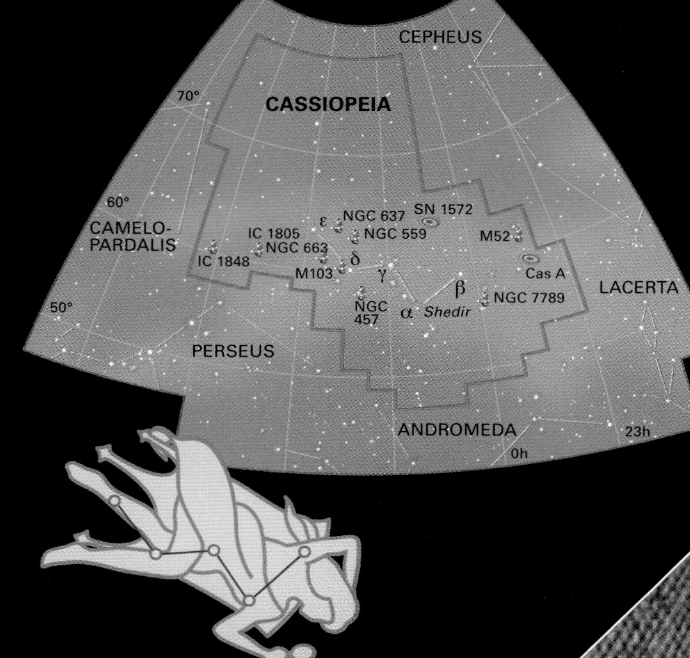

Polar pointer
The centre of the W shape of Cassiopeia points towards the north celestial pole. The constellation is located between Perseus and Cepheus in the Milky Way.

Camelopardalis The Giraffe

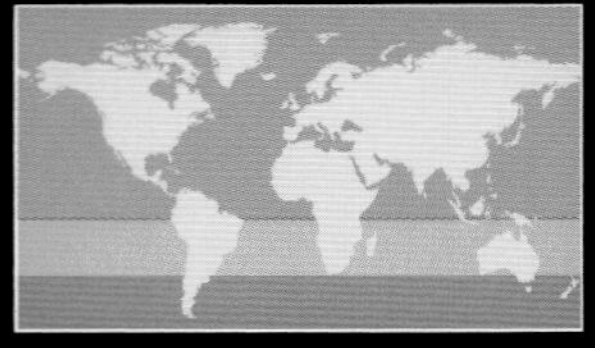

Fully visible 90°N–3°S

This faint constellation of far northern skies was invented in 1613 by Dutch theologian Petrus Plancius. It represents an animal from the Bible.

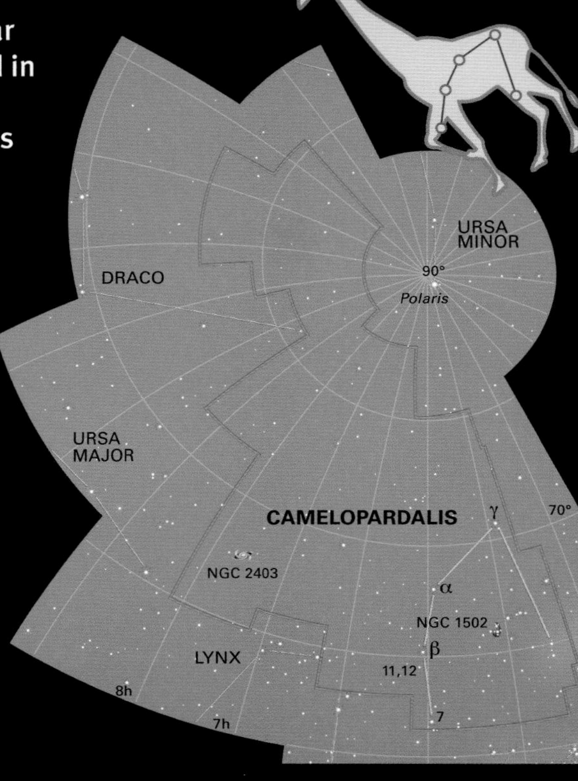

Celestial giraffe
It can be difficult to relate the figure of a giraffe to the stars of Camelopardalis. Only the stars representing the giraffe's legs are visible in this picture.

FEATURES OF INTEREST

Alpha (α) Camelopardalis Despite its designation, alpha is the constellation's second brightest star. It is a blue supergiant, but because it lies about 3,000 light years away, it shines at only magnitude 4.3.

Beta (β) Camelopardalis Outshining alpha with a magnitude of 4.0, beta is a yellow supergiant about 1,000 light years away, with a faint, magnitude-8.6 companion star.

NGC 1502 This small star cluster, with about 45 members, is visible through binoculars and is some 3,100 light years from Earth.

NGC 2403 This spiral galaxy is relatively nearby, at 12 million light years away. A small telescope should show it as an eighth-magnitude, elliptical smudge.

Auriga The Charioteer

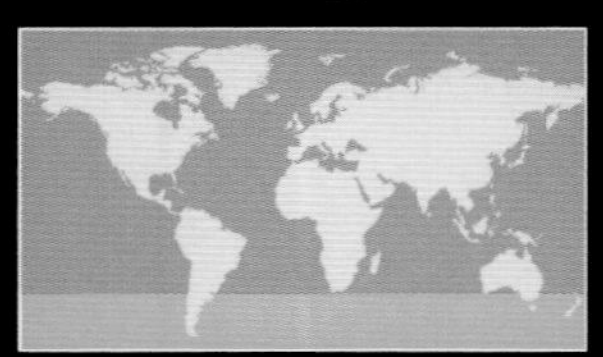

Fully visible 90°N–34°S

This constellation, a highlight of northern winter skies, is usually said to represent Erichthonius, an ancient king of Athens and skilled charioteer. Its southernmost star is shared with Taurus, and the Milky Way passes diagonally across it, making it rich in interesting stars and clusters.

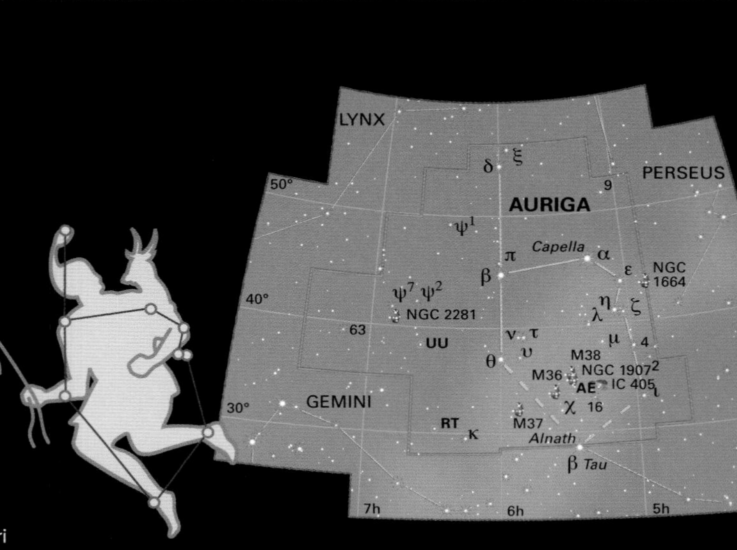

Shared star
Auriga, the charioteer, lies in the Milky Way between Gemini and Perseus. Neighbouring Beta (β) Tauri completes the charioteer figure.

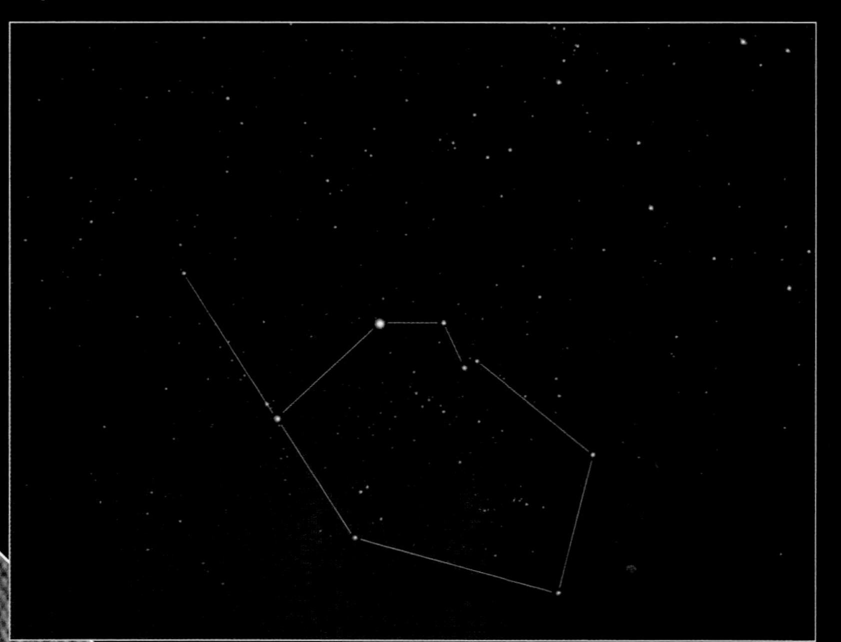

FEATURES OF INTEREST

Alpha (α) Aurigae (Capella) The sixth-brightest star in the sky, Capella shines at magnitude 0.1 and is just 42 light years from Earth. It is actually a binary system, composed of twin yellow giants that orbit each other in 104 days, far too close together for a telescope to separate them.

Zeta (ζ) Aurigae A triangle of stars close to Capella is known as "The Kids". Zeta (ζ), on the southwest point, is an eclipsing binary – a system where two stars passing in front of one another cause regular dips in brightness.

Epsilon (ε) Aurigae The northernmost "kid" is another eclipsing binary, but a very curious one whose eclipses last for about one year out of every 27. This intensely luminous supergiant is orbited by a dark eclipsing partner that seems to be huge and semi-opaque and may be a star with a young, dusty solar system.

Lynx The Lynx

This faint northern group is a relatively late addition to the classical constellations, invented by Johannes Hevelius in the 1680s. A chain of faint stars between Ursa Major and Auriga, it bears no resemblance to the European wildcat of the same name – Hevelius apparently came up with the name because it was so faint that only those with a cat's eyes would be able to spot it.

Fully visible 90°N–28°S

FEATURES OF INTEREST

Alpha (α) Lyncis This is a magnitude-3.2 red giant, 150 light years from Earth.

12 Lyncis To the naked eye, this white star is a faint magnitude 4.9, but a small telescope will reveal a blue-white companion of magnitude 7.3, and a larger instrument will show that the brighter element is itself a binary, making this a triple system, 140 light years from Earth. The components of the brighter star have an orbital period of about 700 years.

NGC 2419 This faint globular cluster, only visible through telescopes of moderate aperture, is 210,000 light years away from Earth, much more distant than the Milky Way's other globular clusters.

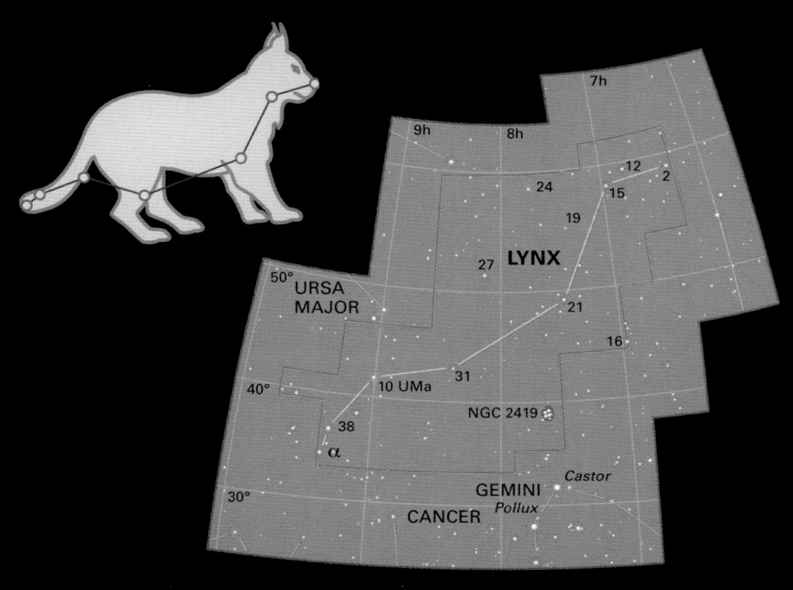

Elusive feline
Lynx consists of nothing more than a few faint stars zigzagging between Ursa Major and Auriga. The constellation has many interesting double and multiple stars.

Ursa Major The Great Bear

One of the best-known constellations of the northern sky, the seven brightest stars in Ursa Major form the familiar pattern of the Plough or Big Dipper – a useful signpost to other stars. But the constellation's fainter stars extend much farther.

Fully visible 90°N–16°S

FEATURES OF INTEREST

Alpha (α) Ursae Majoris (Dubhe) This yellow giant is just over 100 light years away, shining at magnitude 1.8. A line from beta (Merak), through Dubhe points towards the pole star in Ursa Minor.

Zeta (ζ) Ursae Majoris (Mizar) This is a famous double star. Its neighbour Alcor just happens to lie in the same direction, but a small telescope shows that Mizar is also a true binary, with a much closer companion.

M81 This spiral galaxy is bright and just 10 million light years away but can still be seen only through a small telescope.

A familiar sight
The saucepan shape of the Plough's stars is one of the most easily recognized sights in the night sky.

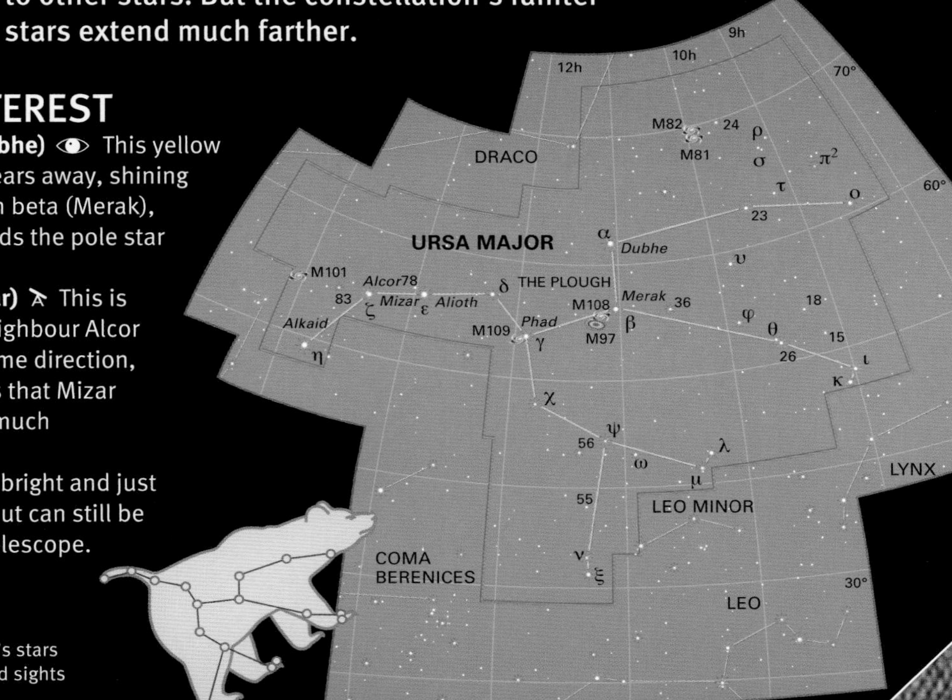

Two bright stars
Canes Venatici represents a pair of hounds, but the unaided eye can see little more than the constellation's brightest stars, Cor Caroli and Beta Canum Venaticorum.

Canes Venatici The Hunting Dogs

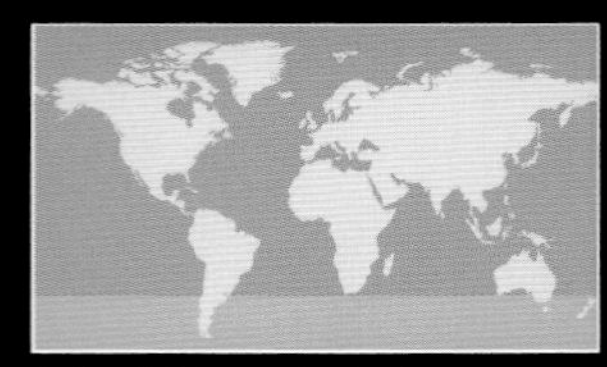

Fully visible 90°N–37°S

This constellation depicts a pair of dogs used by the herdsman Boötes (whose constellation is adjacent) to chase the Great and Little Bears (Ursa Major and Minor) around the north pole. It was formed by Johannes Hevelius at the end of the 17th century.

FEATURES OF INTEREST

Alpha (α) Canum Venaticorum (Cor Caroli) This star's name means "Charles's Heart" – it was named in memory of the executed British King, Charles I. Binoculars will show that it is a wide binary system, composed of white stars with magnitudes 2.9 and 5.6. It lies 82 light years from Earth.

M3 One of the northern sky's best globular clusters, M3 appears as a fuzzy "star" in binoculars and a hazy ball of light through small telescopes.

Whirlpool Galaxy (M51) This spectacular spiral galaxy is bright and relatively close, some 15 million light years away. It happens to lie "face on" to Earth, so binoculars or, ideally, a small telescope will show the bright core, while medium-sized instruments will reveal traces of the spiral arms that give the galaxy its name.

Boötes The Herdsman

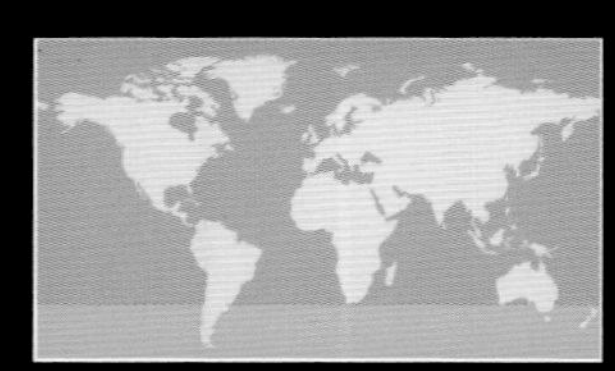

Fully visible 90°N–35°S

The figure of Boötes is often shown driving the Great Bear (see p.61) and Little Bear (see p.58) around the north pole of the sky. Its kite-shaped pattern is quite distinctive. The brightest star in this constellation is Arcturus, which means "bear guard" or "bear-keeper" in Greek.

FEATURES OF INTEREST

Alpa (α) Boötis (Arcturus) Arcturus is one of the closest and brightest stars to us. It is an orange giant nearing the end of its life, and just 36 light years away. At magnitude -0.04, it is the fourth-brightest star in the sky.

Epsilon (ε) Boötis (Izar) Also called Pulcherrima, this is one of the sky's most beautiful double stars. A small telescope will split it to reveal an orange giant of magnitude 2.7 accompanied by a blue star of magnitude 5.1. The pair lie around 150 light years away.

Tau (τ) Boötis This apparently uninspiring star of magnitude 4.5 is notable as host to one of the first planets discovered beyond our solar system. Tau is a yellow star quite similar to the Sun and 51 light years from Earth. A giant planet three times the size of Jupiter orbits it every 3.3 days.

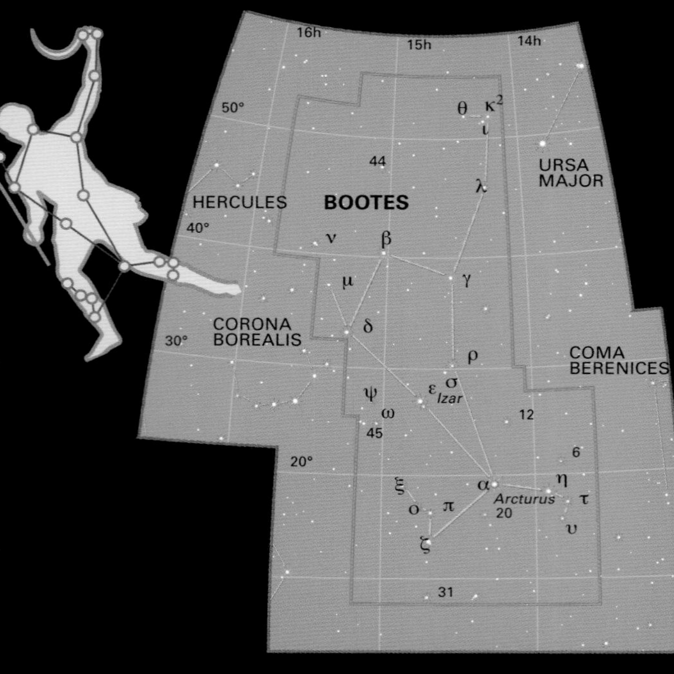

Kite-shaped constellation
Boötes, containing the bright star Arcturus, stands aloft in spring skies in the Northern Hemisphere. This large and conspicuous constellation extends from Draco to Virgo.

Hercules Hercules

Fully visible 90°N–38°S

A large but not particularly prominent constellation depicting the mythical hero and demigod, Hercules can be hard to identify. It is best spotted by working out from the square Keystone at its centre.

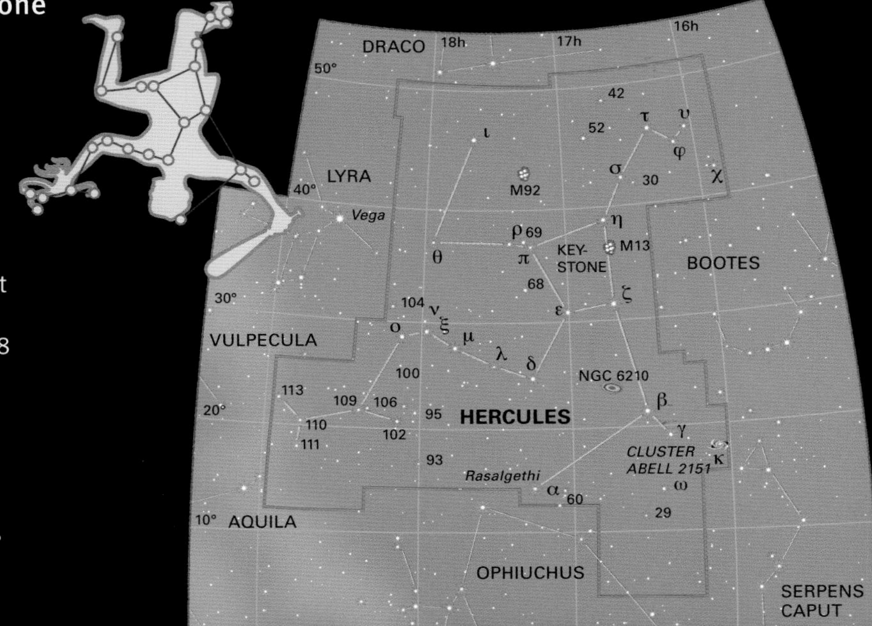

FEATURES OF INTEREST

Alpha (α) Herculis (Rasalgethi) With a name that means "the kneeler's head" in Arabic, this is a double star system containing two stars that orbit one another 380 light years away. One is a huge red giant so large that it has become unstable, and varies in brightness between magnitudes 2.8 and 4.0. The other is a smaller giant shining steadily at magnitude 5.3.

M13 This globular cluster is the best in the northern sky, a knot of 300,000 closely packed stars about 25,000 light years from Earth. Binoculars will show it as a fuzzy ball, and a small telescope should reveal some of the more loosely packed stars around its edges.

pside down
the night sky, Hercules is positioned with his feet pointing ward the pole (top left) and his head pointing south. He eels with one foot on the head of the celestial dragon Draco.

Lyra The Lyre

lly visible 90°N–42°S

Although it is one of the smaller constellations, Lyra is easily spotted in northern skies, thanks to the presence of the fifth-brightest star in the sky, brilliant white Vega. It represents the ancient musical instrument played by Orpheus on his journey to the underworld.

EATURES OF INTEREST

lpha (α) Lyrae (Vega) This is a white star st 25 light years from Earth. It shines at agnitude 0.0, indicating that it is some ɔ times as luminous as the Sun, and it is ırrounded by an intriguing disc of dusty ebris that may be left over from the formation f a planetary system.

psilon (ε) Lyrae This famous multiple star plits into a double when viewed through inoculars, but a small telescope will show that ach of these components is itself also double, naking epsilon a "double double" system.

ne Ring Nebula (M57) Lyra's other nowpiece is the Ring Nebula, the most famous lanetary nebula in the sky. Lying midway etween beta and gamma, M57 is a delicate hell of gas cast off by a dying star 1,100 light ears away. It shines at magnitude 9.5, and is est seen through a small telescope.

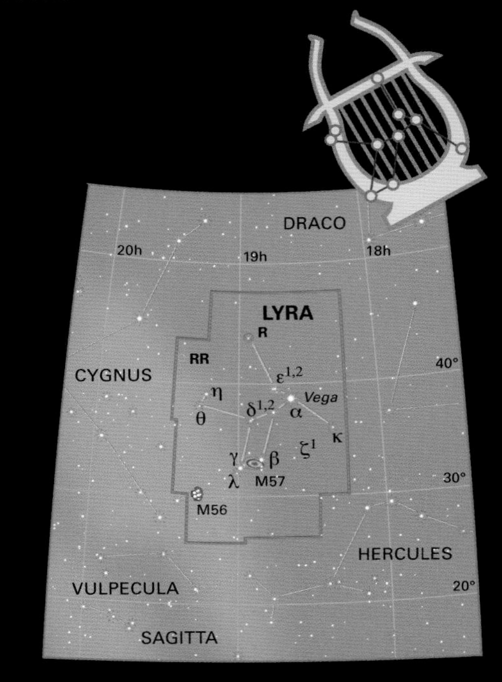

Stringed instrument
Lyra, dominated by dazzling Vega, represents the harp played by Orpheus, the musician of Greek myth. Arab astronomers visualized the constellation as an eagle or vulture. It lies on the edge of the Milky Way, next to Cygnus.

Cygnus The Swan

Fully visible 90°N–28°S

Sometimes known as the Northern Cross because of its distinctive shape, Cygnus represents a swan flying down the Milky Way. As well as rich starfields, it contains many deep sky objects of interest.

FEATURES OF INTEREST

Alpha (α) Cygni (Deneb) Although at magnitude 1.3 it is outshone by nearby Vega, Deneb is one of the sky's most luminous stars – it lies 2,600 light years from Earth, and must be some 160,000 times brighter than the Sun.

Beta (β) Cygni (Albireo) This is a beautiful double star of contrasting colours – binoculars will split it into yellow and blue stars of magnitudes 3.1 and 4.7.

Cygnus Rift Cygnus contains several interesting nebulae, but the most obvious is this dark cloud of gas and dust that runs alongside the swan's neck and obscures our view of the Milky Way behind it.

Cygnus X-1 Although undetectable to amateurs, this strong X-ray source is thought to mark the site where a black hole is pulling material from a companion star.

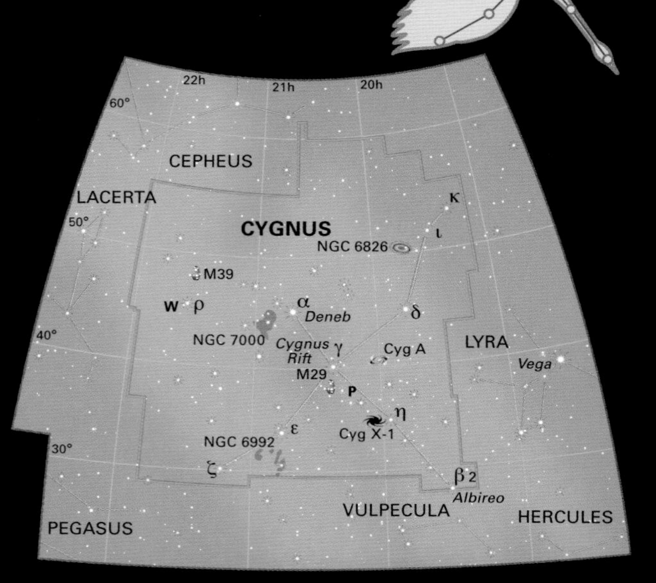

Poised in flight
One of the most prominent constellations of the northern sky, Cygnus depicts a swan flying with its wings outstretched. The beak of the swan is marked by a double star, Beta (β) Cygni.

Andromeda Andromeda

Fully visible 90°N–37°S

Easy to find because of its link to the Square of Pegasus, Andromeda – daughter of Cassiopeia (see p.59) – represents a princess chained to a rock as a sacrifice to the sea monster Cetus, but ultimately rescued by the hero Perseus.

FEATURES OF INTEREST

Alpha (α) Andromedae (Alpheratz) Sometimes also referred to as Delta Pegasi, Alpheratz is a blue-white star 97 light years from Earth.

Gamma (γ) Andromedae (Almach) Through small telescopes, Almach appears as a contrasting double, with yellow and blue stars of magnitudes 2.3 and 4.8. Larger telescopes will also show the blue star's fainter, sixth-magnitude companion.

The Andromeda Galaxy (M31) The most distant object visible to the naked eye, this looks at first like a fuzzy, fourth-magnitude star. Binoculars or a small telescope reveal an elliptical disc – the bright central area of a huge spiral galaxy larger than the Milky Way, 2.5 million light years away. Two companion galaxies are visible through small telescopes.

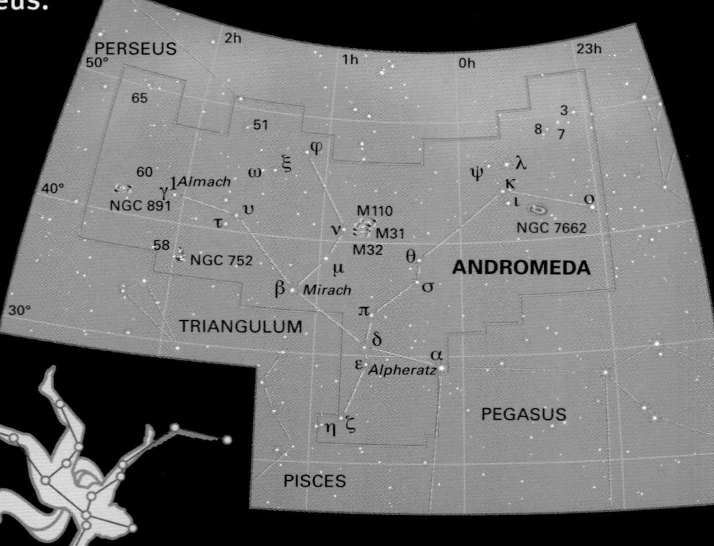

Head to toe
Andromeda is one of the original Greek constellations. Its brightest stars represent the princess's head (α), her pelvis (β), and her left foot (γ).

Lacerta The Lizard

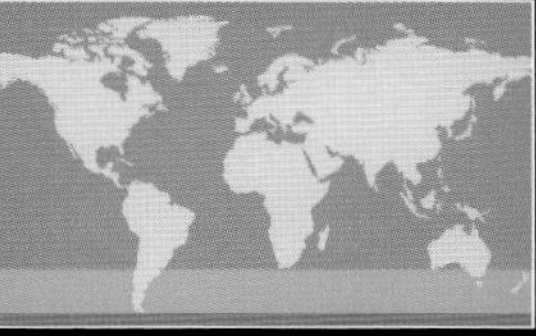

ully visible 90°N–33°S

Straddling the northern Milky Way between Cassiopeia and Cygnus, Lacerta is a small and obscure constellation introduced by the Polish astronomer Johannes Hevelius in 1687. Its size means that it contains few significant deep-sky objects, but it is the site of occasional nova explosions (produced when a star brightens suddenly). It also contains the prototype of a very strange class of galaxy.

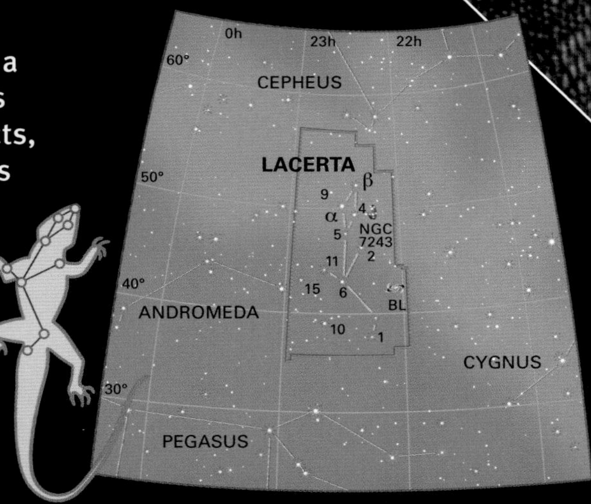

FEATURES OF INTEREST

lpha (α) Lacertae This blue-white star shines at nagnitude 3.8 and lies 102 light years from Earth. This neans that it is roughly 27 times as luminous as the Sun.

IGC 7243 This loose group of blue-white stars, thought o lie about 2,800 light years from Earth, is so scattered that ome astronomers suspect it is not a true open cluster at all.

BL Lacertae This strange and rapidly varying starlike object is in fact a "blazar" – a distant galaxy with a massive black hole at its centre, that is gulping down material from its surroundings and spitting it out in a jet that points directly at Earth. Because we see these jets head-on from Earth, they tend to look starlike.

Triangulum The Triangle

ully visible 90°N–52°S

This small northern constellation fills the gap between Perseus, Andromeda, and Aries. Despite its lack of bright stars, it has an ancient origin – Greek astronomers originally saw it as a version of their letter "delta". Its compact size makes it relatively easy to spot nevertheless.

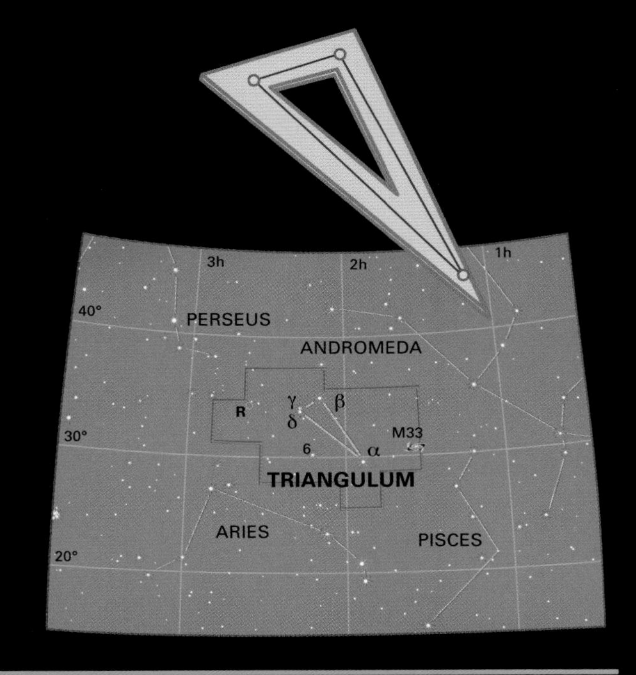

FEATURES OF INTEREST

lpha (α) Trianguli With a magnitude of 3.4, this is a vhite star some 65 light years away. Although designated lpha, it is not in fact Triangulum's brightest star.

eta (β) Trianguli Alpha's brighter neighbour shines t magnitude 3.0 and lies about 135 light years from Earth. Despite the stars' apparent similarity, beta must be considerably more luminous in order to outshine alpha – it s classed as a "giant". Apart from these two stars and M33, here is little else of note in this constellation.

M33 Triangulum's finest sight is this spiral galaxy, one of the closest in the sky at 2.7 million light years away. In spite of its size and proximity, it is hard to spot in binoculars or a small telescope because it presents its "face" to Earth, and so its light is thinly spread. M33, also known as the Triangulum Galaxy, is the third major member of our Local Group of galaxies, after the Andromeda Galaxy (M31) and the Milky Way itself. In long-exposure photographs, it looks like a starfish. It may actually be in orbit around Andromeda.

Perseus Perseus

ully visible 90°N–31°S

This constellation represents a hero of Greek myth, coming to the rescue of nearby Andromeda. He carries the head of Medusa, a creature whose gaze could turn anyone to stone. It is one of the original Greek constellations.

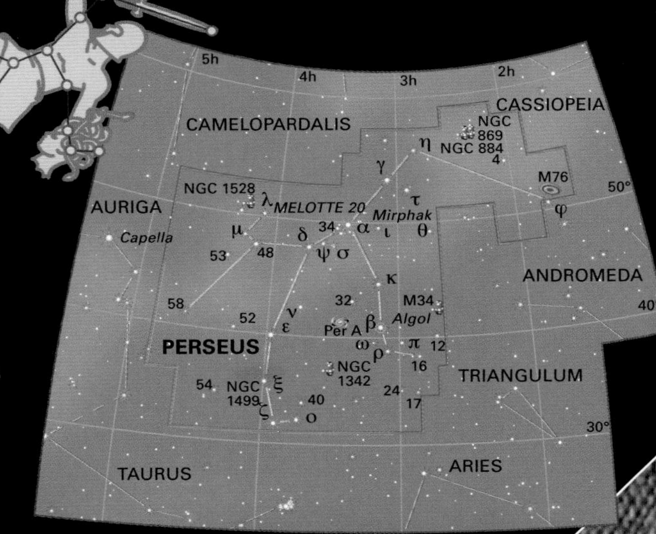

FEATURES OF INTEREST

lpha (α) Persei (Mirphak) Binoculars will reveal that this magnitude-1.8 yellow supergiant lies at the heart of a cluster of fainter blue stars, 590 light years from Earth.

eta (β) Persei (Algol) This famous variable star is also nown as the "winking demon". It was the first eclipsing inary to be identified – a double star with two components hat pass in front of each other every 2.87 days, causing the star's apparent brightness to dip from magnitude 2.1 to 3.4 for about 10 hours.

Double Cluster (NGC 869, NGC 884) This famous cluster is a spectacular sight in binoculars and can be seen with the naked eye as a bright knot in the Milky Way. The clusters are both about 7,000 light years from Earth, and genuine neighbours in space.

Aries The Ram

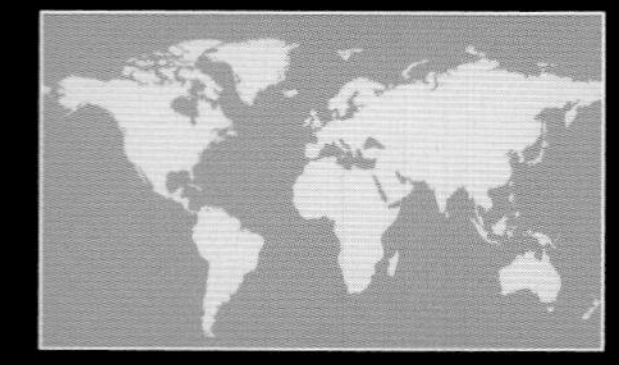

Fully visible 90°N–58°S

The zodiac constellation of Aries represents the ram with the golden fleece in the legend of Jason and the Argonauts. Although it has astronomical and astrological significance as the original site of the "First Point of Aries" (the point where the ecliptic crosses the celestial equator, defining zero hours right ascension), this point now lies in the neighbouring constellation Pisces. Aries's pattern is relatively faint and hard to identify.

Legendary ram
From a crooked line formed by three faint stars, ancient astronomers visualized the figure of a crouching ram, with its head turned back over its shoulder.

FEATURES OF INTEREST

Alpha (α) Arietis (Hamal) This yellow giant star, about 66 light years from Earth, shines at magnitude 2.0. Its popular name is derived from the Arabic for "lamb".

Gamma (γ) Arietis (Mesartim) This is an attractive double star – one of the first discovered to be double. It was found by English scientist Robert Hooke in 1664. Small telescopes will easily separate it to reveal twin white components of magnitude 4.8. The stars are orbiting each other some 200 light years from Earth.

Lambda (λ) Arietis This is another double star – binoculars will reveal that the white, magnitude-4.8 primary star has a yellow companion of magnitude 7.3.

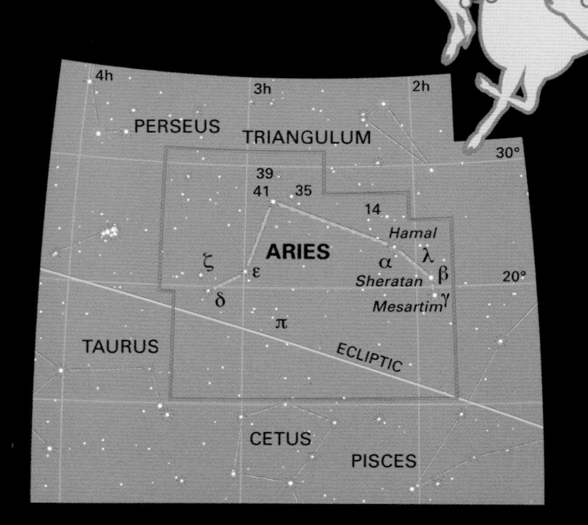

Taurus The Bull

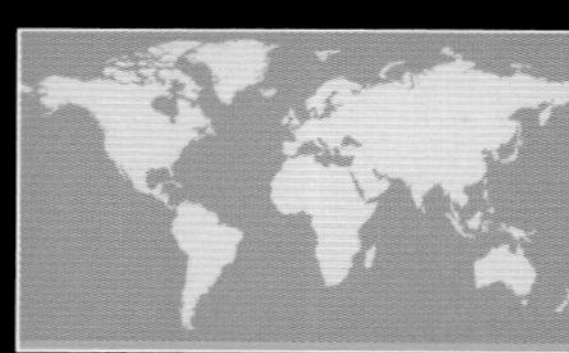

Fully visible 88°N–58°S

This rich constellation represents a bull charging the hunter Orion. One of the oldest constellations, it has been recognized since Babylonian times – perhaps because of the distinctive "face" formed by Aldebaran and the Hyades.

FEATURES OF INTEREST

Alpha (α) Tauri (Aldebaran) This red giant star lies some 65 light years from Earth. It shines at around magnitude 1.0, but its brightness varies because this elderly star has become unstable.

The Hyades This V-shaped star cluster lies well beyond Aldebaran, some 160 light years away. Binoculars will reveal stunning starscapes.

The Pleiades (M45) Named after a group of mythical Greek nymphs, this famous open cluster marks the bull's shoulders. Naked-eye observers usually see six of the so-called "seven sisters", but binoculars or a telescope show many more hot blue stars. The cluster is just 50 million years old and lies 400 light years away.

Crab Nebula (M1) This nebula is the shredded remnant of a star that exploded as a supernova in 1054.

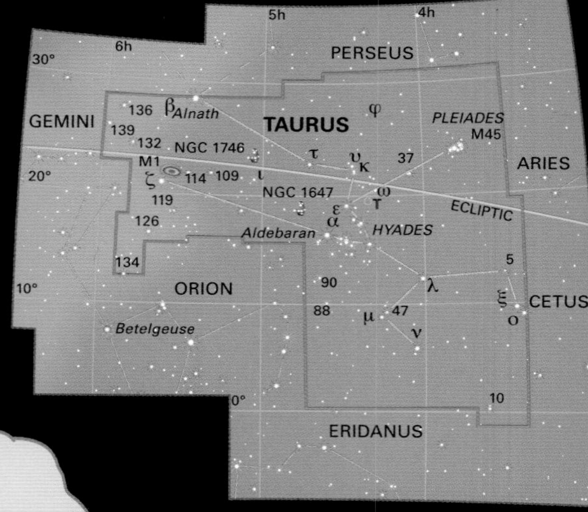

Raging bull
Taurus, the celestial bull, thrusts his star-tipped horns into the night air. The bull is said to represent a disguise adopted by Zeus in a Greek myth.

Gemini The Twins

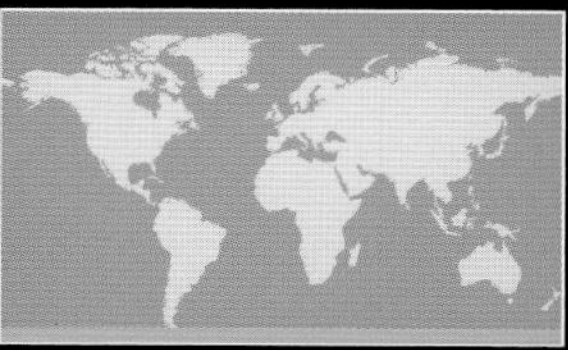

Fully visible 90°N–55°S

This zodiac constellation represents the twins Castor and Pollux, who were the brothers of Helen of Troy and were among the crew of the Argo that went in search of the golden fleece. Gemini is easy to find due to its proximity to Orion.

FEATURES OF INTEREST

Alpha (α) Geminorum (Castor) Castor is a fascinating multiple star system with overall magnitude 1.6. A small telescope will divide it into two white stars, while a larger one reveals a faint red companion. Each of these stars is itself a double (though none can be separated visually), giving Castor a total of six stars.

Beta (β) Geminorum (Pollux) In contrast to Castor, Pollux is a single yellow star, about 34 light years from Earth. It is also brighter than Castor, at magnitude 1.1.

M35 This open cluster can be spotted with the naked eye and is a rich target for binoculars, through which it appears as an elongated, elliptical patch of starlight spanning the same apparent width as a full moon.

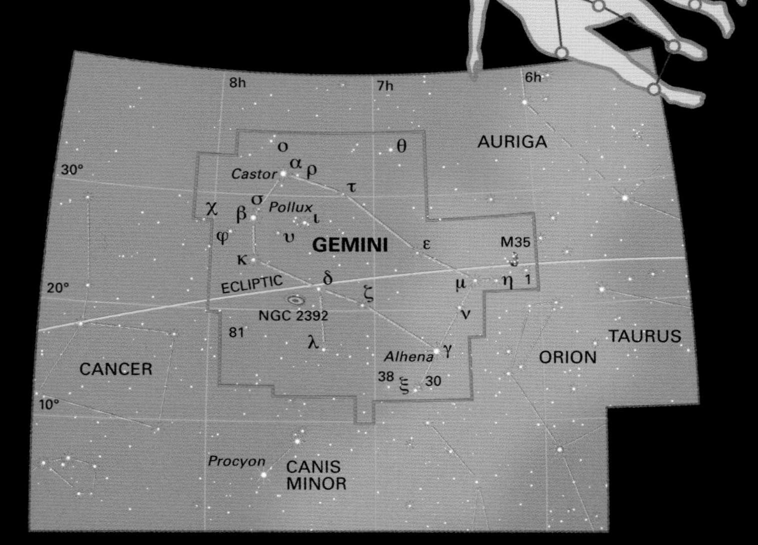

Celestial twins
Castor and Pollux, the twins of the Greek myth, stand side by side in the sky between Taurus and Cancer. The bright "star" in the middle of Gemini is actually the planet Saturn.

Cancer The Crab

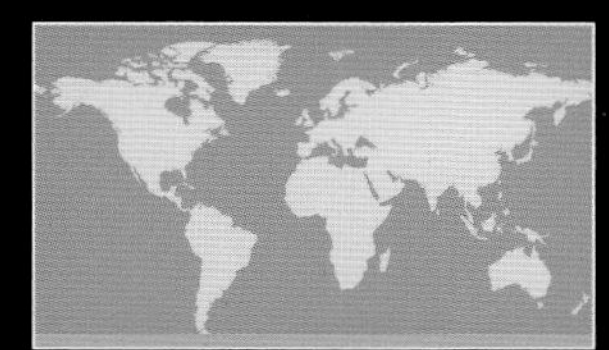

Fully visible 90°N–57°S

Although Cancer's star pattern is faint and indistinct, it is still easy to find. This is because it lies between the brighter stars of Leo and Gemini. Cancer represents a crab that attacked the hero Hercules, but was crushed beneath his foot.

FEATURES OF INTEREST

Alpha (α) Cancri (Acubens) This star, whose name means "the claw", is actually fainter than nearby beta. It is a white star of magnitude 4.2, some 175 light years from Earth.

Beta (β) Cancri (Altarf) Cancer's brightest star is an orange giant, 290 light years from Earth and obviously brighter than Acubens at magnitude 3.5.

Praesepe (M44) This is a group of 50 young stars spread across an area of sky three times the size of the full Moon. Although their combined light makes them easily visible to the naked eye, binoculars are needed to resolve the individual stars. This cluster is also known as the Beehive. Its traditional name, Praesepe, means "manger".

Hidden crab
Cancer is the faintest constellation in the zodiac, but it contains a major star cluster, M44, which is just visible near the centre of the constellation.

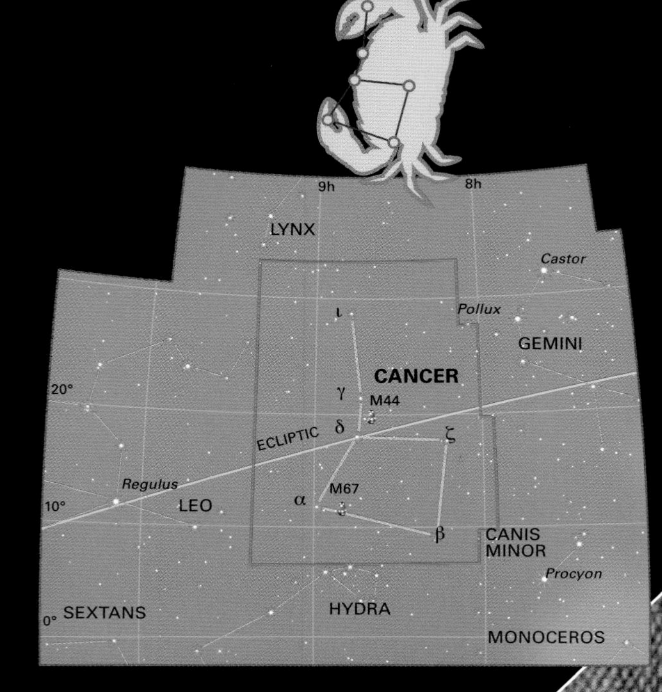

Leo Minor The Lesser Lion

Fully visible 90°N–48°S

This constellation was invented by Polish astronomer Johannes Hevelius around 1680. He claimed that its stars resembled nearby Leo, but the resemblance is far from obvious, and it seems as though Hevelius was simply keen to fill a gap in the sky for his great star atlas, *Uranographia*.

FEATURES OF INTEREST

46 Leonis Minoris The constellation's brightest star is an orange giant of magnitude 3.8. It lies about 80 light years from Earth and is nearing the end of its life. It has missed out on a Greek letter designation due to obscurity and historical accident. The 19th-century English astronomer Francis Baily overlooked recording this star as alpha (α).

Beta (β) Leonis Minoris The second brightest star in the constellation, meanwhile, does merit a Greek letter. This yellow giant shines at magnitude 4.2 and is 190 light years away, so in reality it is considerably more luminous than 46.

R Leonis Minoris Lying just to the west of 21 Leonis Minoris, this is an interesting star to track with binoculars. It is a pulsating red giant with a period of 372 days, similar to the famous Mira in Cetus. At its peak, of magnitude 6.3, it is easily spotted in binoculars, but at its dimmest it fades beyond the reach of small telescopes.

Coma Berenices Berenice's Hair

Fully visible 90°N–56°S

This constellation (whose name is often shortened to simply Coma) represents the hair of a mythical queen of Egypt. Despite a lack of bright stars, it is easily located since it lies between the brighter stars of Leo and Boötes. It contains significant clusters of both stars and galaxies.

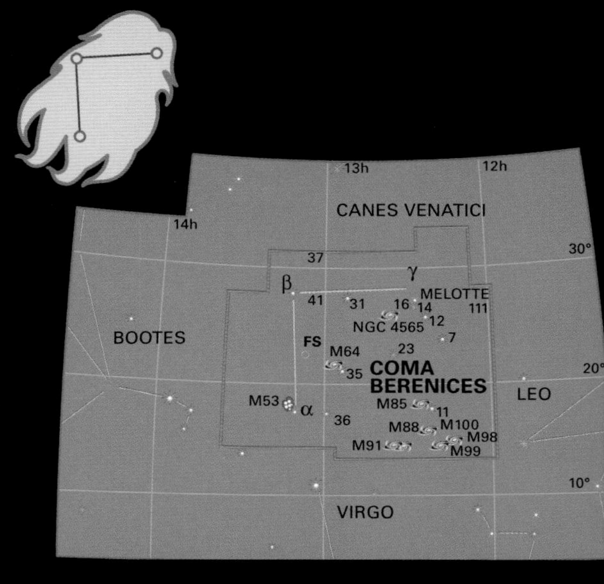

FEATURES OF INTEREST

Melotte 111 The constellation gets its name from the faint strands of stars in this open cluster. Melotte 111 is one of the closest open star clusters to Earth, and more than 20 of its stars are visible to the naked eye.

M53 The brighter of two globular clusters in Coma, M53 is some 56,000 light years away, visible through binoculars but best seen in a small telescope.

Coma Cluster Many galaxies are scattered across this part of the sky. Some are overspill from the Virgo Cluster, around 50 million light years away, while others are members of the more distant Coma Cluster, centred around a point close to beta.

M64 Coma's brightest galaxy is nicknamed the "Black Eye". It is a spiral tilted at an angle to the Earth, and is crossed by a prominent lane of dust.

Leo The Lion

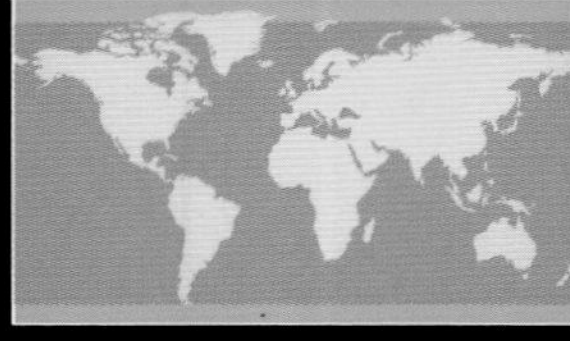

Fully visible 82°N–57°S

This zodiac constellation, which represents the Nemean lion fought by Hercules, does bear some resemblance to a resting lion. The Leonid meteors radiate from the region around the head and neck, which is called the Sickle, every November.

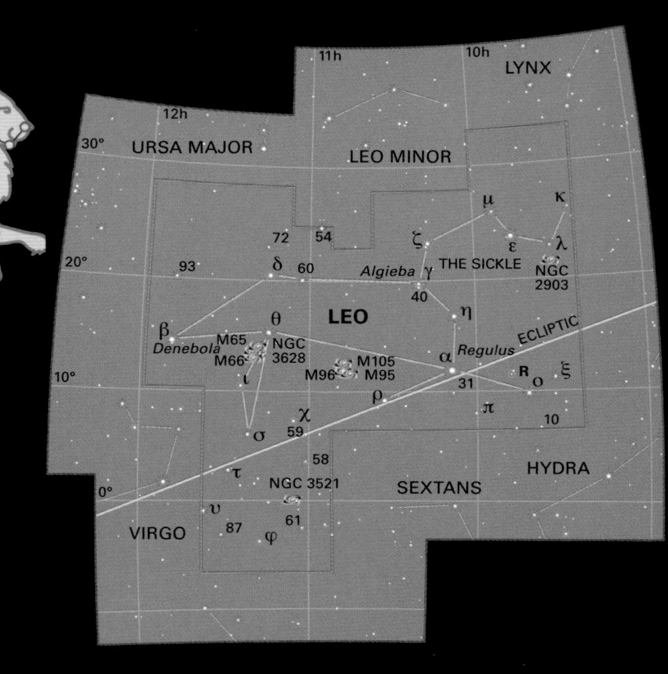

FEATURES OF INTEREST

Alpha (α) Leonis (Regulus) This bright blue-white star shines at magnitude 1.4 and is located almost 80 light years away. It lies at the foot of the pattern of six stars marking the lion's head and chest. It has a companion star of magnitude 7.8 that can be seen through binoculars.

Gamma (γ) Leonis (Algieba) This attractive double star consists of two yellow giants some 170 light years form Earth. Easily split in small telescopes, the brighter star is magnitude 2.0 and the fainter 3.2. Both stars orbit each other every 600 years or so.

R Leonis This red giant is 3,000 light years away. It varies in brightness over 312 days, usually staying below naked-eye visibility but peaking at fourth magnitude.

Virgo The Virgin

Often associated with the harvest goddess Demeter, Virgo lies to the southeast of the more identifiable Leo. Demeter is usually depicted holding an ear of wheat, which is represented by Spica, the constellation's brightest star. Virgo is also identified as Dike, the Greek goddess of justice. It is host to Earth's nearest major galaxy cluster.

Fully visible 67°N–75°S

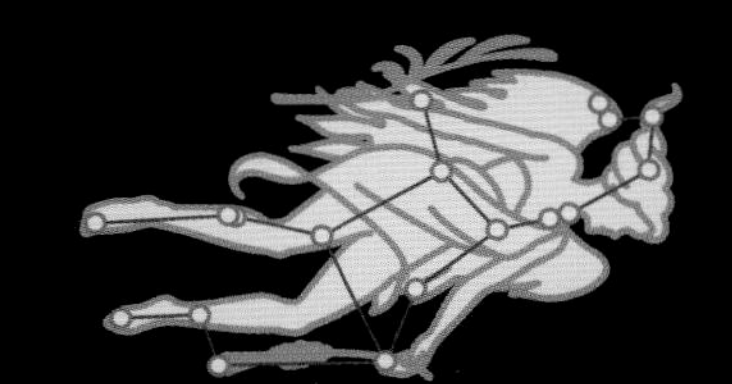

The virgin goddess constellation
Virgo straddles the celestial equator, between Leo and Libra. It is the largest constellation of the zodiac, and the second-largest overall.

FEATURES OF INTEREST

Alpha (α) Virginis (Spica) This bright star has an average magnitude of 1.0. About 260 light years from Earth, it is actually a binary star – its elements cannot be split visually, but the companion distorts the primary's shape, causing its brightness to vary as different amounts of the surface are presented to Earth.

M87 This huge "giant elliptical" galaxy is at the heart of the Virgo Cluster of galaxies. It shines at magnitude 8.1 and lies 50 million light years away.

Sombrero Galaxy (M104) This bright galaxy is 35 million light years away, and is considerably closer than the Virgo Cluster. It is an edge-on spiral that appears Saturn-like, and a dark lane of dust in the galaxy's plane crosses its central bulge. Only its nucleus is visible through small telescopes. The dust lane is only revealed when seen through a large-aperture telescope or on long-exposure photographs.

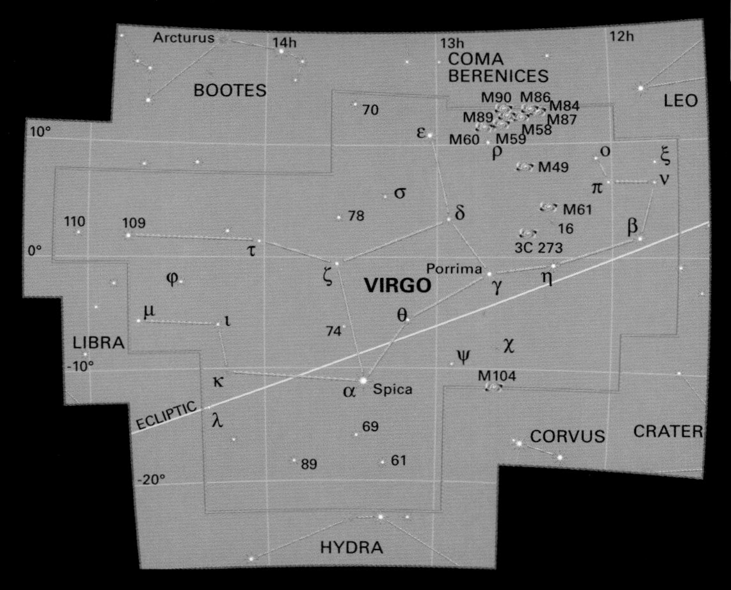

Libra The Scales

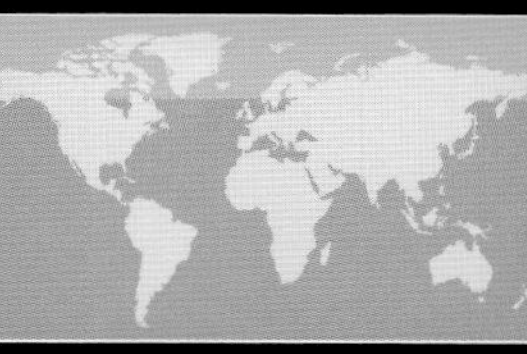

The only sign of the zodiac depicting an object rather than a living creature, Libra was once seen by the ancient Greeks as Chelae Scorpionis, the claws of neighbouring Scorpius. Since Roman times, Libra has been interpreted as the scales of justice, held by nearby Virgo.

Fully visible 60°N–90°S

FEATURES OF INTEREST

Alpha (α) Librae (Zubenelgenubi) With a name meaning "the southern claw" in Arabic, alpha is a bright double easily split with binoculars or even with sharp unaided eyesight. Its two stars, a blue-white giant of magnitude 2.8 and a white star of magnitude 5.2, lie 70 light years from Earth. To the north of this pair is the constellation's brightest star, Zubeneschamali, meaning "the northern claw", or beta (β) Librae.

Mu (μ) Librae This double star, with components of magnitudes 5.6 and 6.7, is a close pairing 235 light years from Earth, but can be split by all but the smallest telescopes.

48 Librae Lying 510 light years from Earth, this young star is at an early stage of its development, and is still throwing off excess material that forms shells around the star, causing it to vary unpredictably by around 0.1 magnitudes.

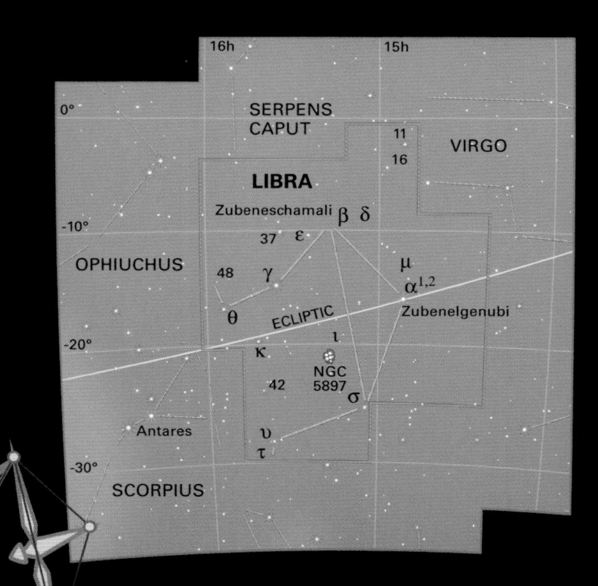

Libra's stars
Originally, the ancient Greeks visualized Libra as the claws of Scorpius, which is why the constellation's brightest stars have names meaning "northern claw" and "southern claw".

Starry crown
Like a celestial tiara, the seven main stars of Corona Borealis form an arc between Bootes and Hercules. According to the myth, Dionysus threw Ariadne's jewelled crown into the sky, where it transformed into stars.

Corona Borealis The Northern Crown

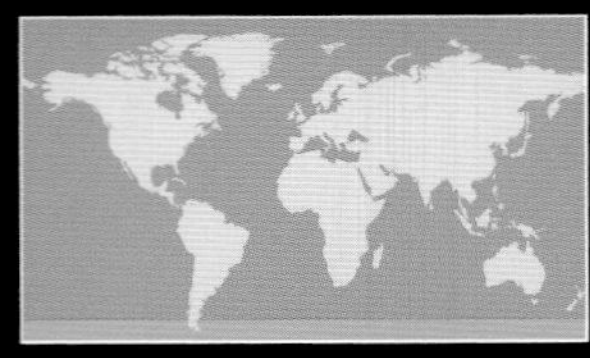

Fully visible 90°N–50°S

The distinctive arc of Corona Borealis lies just to the east of Boötes, and is easily spotted despite the relative faintness of its stars. It contains a number of interesting variable stars. The constellation depicts a crown worn in Greek myth by Princess Ariadne at her wedding to the god Dionysus.

FEATURES OF INTEREST

Alpha (α) Coronae Borealis (Alphekka) 👁 This is an eclipsing binary star, similar to Algol in Perseus, although far less obvious, since it varies by only 0.1 magnitude either side of its average of 2.2 in a 17.4-day cycle.

R Coronae Borealis 👁 This intriguing variable is normally just visible to the naked eye, enclosed by the curve of the crown, shining at magnitude 5.8. But every few years it unpredictably plunges in brightness, disappearing beyond the range of most amateur telescopes. R is a yellow supergiant some 6,000 light years from Earth and seems to throw off shells of material that obscure its own light.

T Coronae Borealis 🔭 Every few decades this nova system, known as the Blaze Star, does the opposite of R, brightening rapidly from magnitude 11 to around 2.

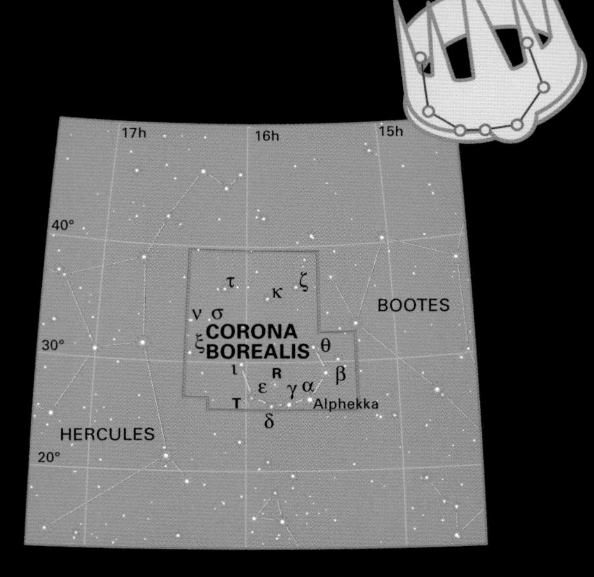

Serpens The Serpent

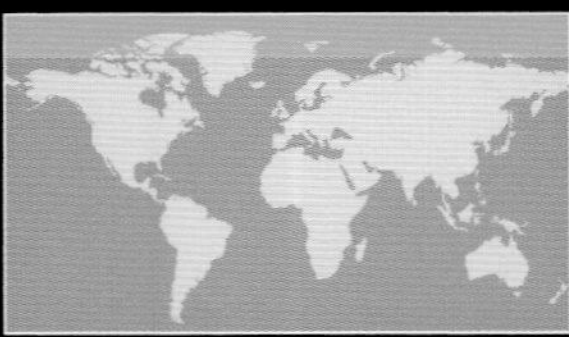

Fully visible 74°N–64°S

This is one of the 48 original Greek constellations, representing a snake coiled around Ophiuchus. Uniquely, it is split into two parts, on either side of Ophiuchus – Serpens Caput represents the snake's head, and Serpens Cauda its tail. Both parts straddle the celestial equator.

FEATURES OF INTEREST

Alpha (α) Serpentis (Unukalhai) 👁 Situated in Serpens Caput, this is an orange giant of magnitude 2.7, 70 light years away.

M5 👁 An attractive globular cluster, M5 is just visible to the naked eye on dark nights, hovering around magnitude 5.6. Binoculars or a small telescope reveal a hazy ball of light, but at 24,500 light years away, larger telescopes are needed to see the curving chain of individual stars.

M16 🔭 This open cluster of about 60 stars lies 8,000 light years away at the heart of the large, faint Eagle Nebula, a huge cloud of gas and dust from which the stars have recently been born. It appears as a hazy patch covering an area of sky similar in size to a full moon.

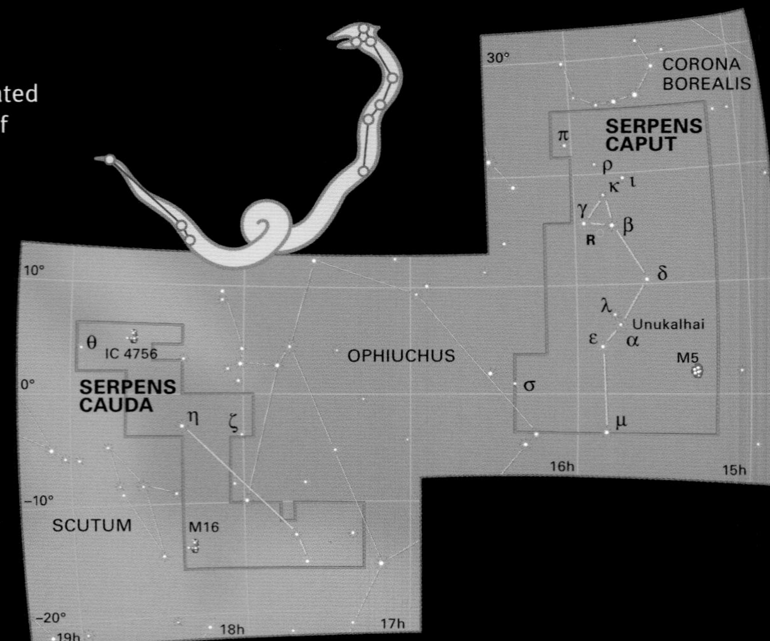

Serpentine stars
The upper part of the snake contains Unukalhai (α), which derives its name from the Arabic for "the serpent's neck". In Greek mythology, snakes were a symbol of rebirth, because of their ability to shed their skin.

Ophiuchus The Serpent Holder

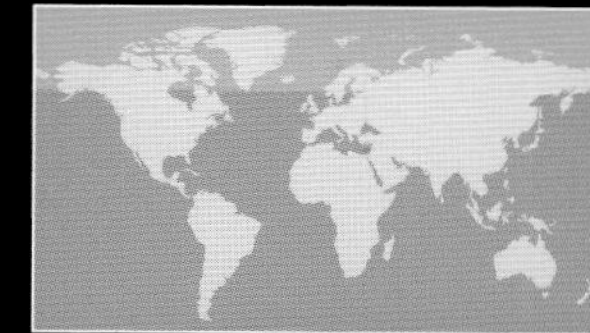

Fully visible 59°N–75°S

This large, indistinct constellation is either represented as Hercules wrestling the snake Serpens or alternatively as Asclepius, the Greek god of healing, who carried a staff with a serpent entwined around it. The serpent is represented by adjoining Serpens.

FEATURES OF INTEREST

Alpha (α) Ophiuchi (Rasalhague) Ophiuchus's brightest star is this magnitude-2.1 white giant, about 50 light years away.

Rho (ρ) Ophiuchi This fine multiple star is still embedded in the faint gas from which it formed. Binoculars will show two wide companions close to the magnitude-5.0 primary star, while a small telescope will show that the primary has a closer neighbour of magnitude 5.9.

Barnard's Star Though too faint for binoculars, this star (found near beta) is interesting as it is the fastest-moving in the sky. It is just six light years from Earth, and moving so fast that it crosses a Moon's width of the sky every 200 years. This celebrated star is the second-closest star to the Sun.

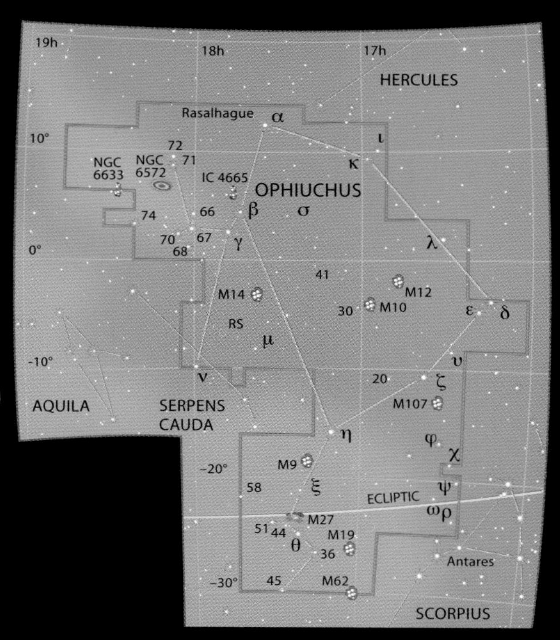

Snake man
Ophiuchus represents a man wrapped in the coils of a huge snake, the constellation Serpens. The ecliptic runs through Ophiuchus, and planets can be seen within its borders.

Scutum The Shield

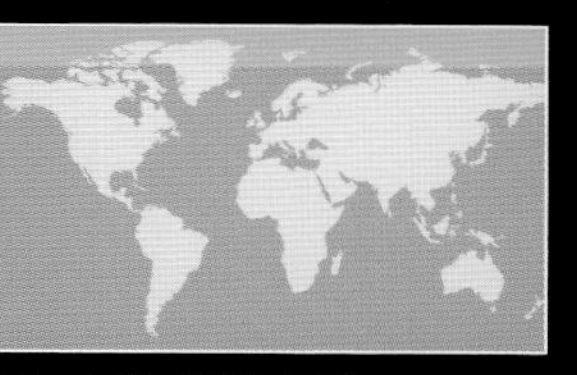

Fully visible 74°N–90°S

This small, kite-shaped constellation was invented in the 17th century by the Polish astronomer Johannes Hevelius and was originally named Scutum Sobieski (meaning "Sobieski's shield") in honour of Hevelius's patron, the King of Poland. Lying in the Milky Way, it is best located by searching between Altair (in the neighbouring constellation Aquila) and the bright stars of Sagittarius.

FEATURES OF INTEREST

Delta (δ) Scuti This magnitude-4.7 star, 160 light years from Earth, is the prototype for a class of rapidly changing variable stars, though it only varies by about 0.1 magnitude over each 4.6-hour cycle.

R Scuti This is a variable star with a slower period than Delta Scuti, and its changes are much easier to follow. It is a yellow supergiant that varies from an easy magnitude 4.5 at its peak, down to magnitude 8.8, the limit of binocular visibility, in a cycle lasting 144 days.

Wild Duck Cluster (M11) A rich open cluster, this is easy to see with the naked eye and is rewarding for binoculars. When seen through a telescope, the stars form a fan shape, like a flock of ducks in flight, hence the popular name.

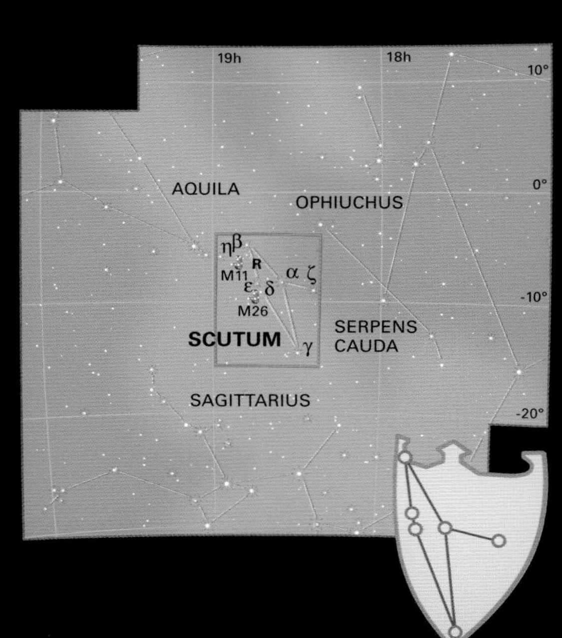

Sobieski's shield
Scutum is a minor constellation with no bright stars of its own, but it lies in an area of the Milky Way, between Aquila and Sagittarius, that is particularly rich with stars. One of the brightest parts of the Milky Way lies in Scutum and is known as the Scutum Star Cloud.

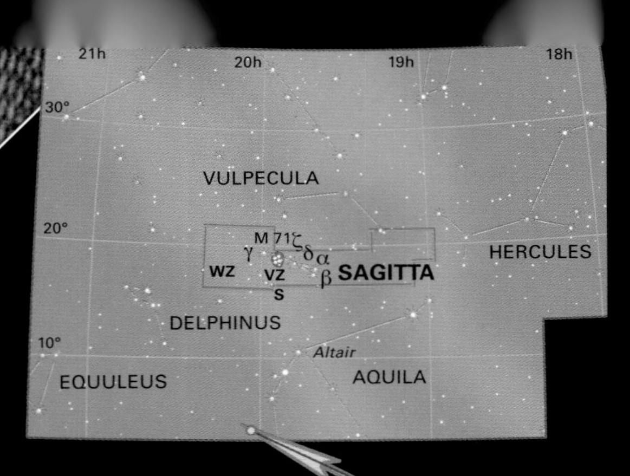

Fully visible 90°N–69°S

Although this small constellation represents an arrow, and was recognized in ancient times, it has nothing to do with the larger constellation of Sagittarius, the archer. Sagitta lies in the Milky Way and represents an arrow fired by Hercules towards Aquila and Cygnus.

FEATURES OF INTEREST

Alpha (α) (Sham) and Beta (β) Sagittae Alpha and beta are twin yellow stars, both of magnitude 4.4. They are genuine neighbours in space, both lying about 470 light years away. Alpha's Arabic name, Sham, means "arrow".

Gamma (γ) Sagittae The brightest star in Sagitta, gamma is an orange giant of magnitude 3.5. It lies at a distance of 175 light years from Earth, at the tip of the arrow that points northeastward.

S Sagittae This yellow supergiant, 4,300 light years from Earth, is a pulsating variable, ranging in brightness between magnitudes 5.5 and 6.2 every 8.38 days.

M71 This star cluster is usually classed as a globular, but its relatively loose structure means that some astronomers suspect it is really a large open cluster, as it lacks the central condensation typical of globulars. It lies about 13,000 light years away, and is visible only through binoculars.

Aquila The Eagle

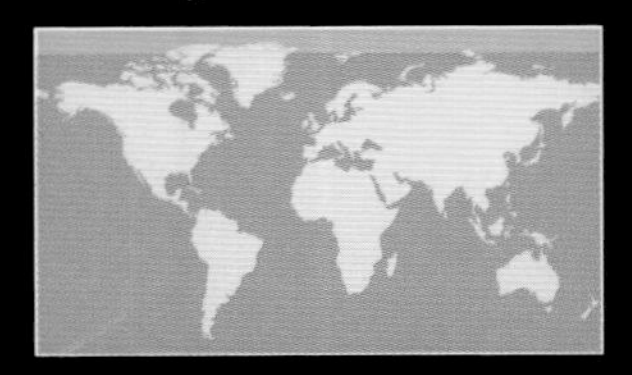

This constellation is found easily by its central bright star Altair, flanked by twins. Aquila depicts the god Zeus, who took the form of an eagle to carry away the youth Ganymede (himself depicted by Aquarius).

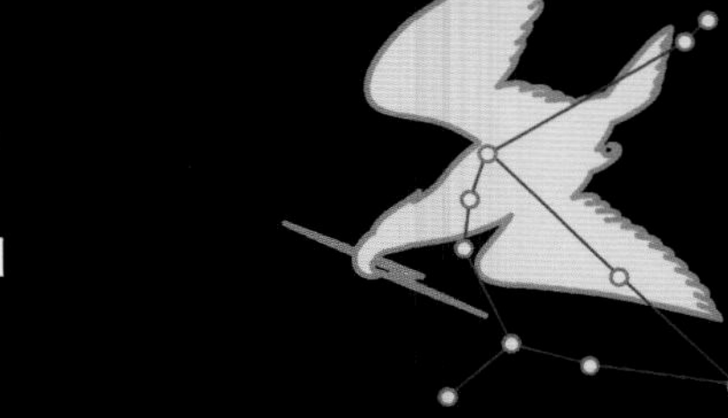

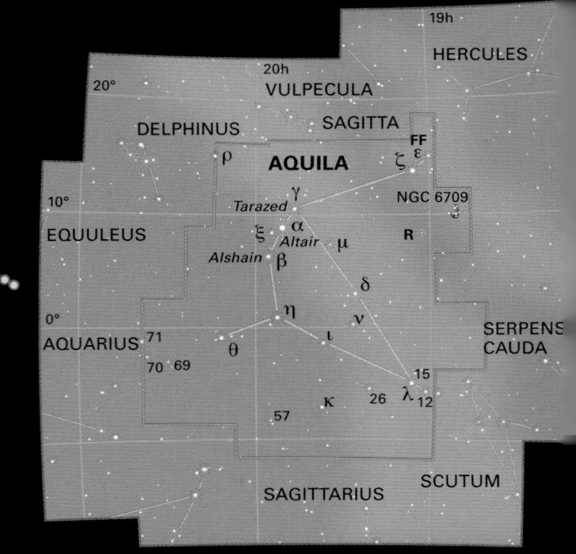

FEATURES OF INTEREST

Alpha (α) Aquilae (Altair) Altair is one of the closest bright stars to Earth, lying just 17 light years away and shining at magnitude 0.8. Along with Deneb (in Cygnus) and Vega (in Lyra), it forms the "summer triangle" of northern skies.

Beta (β) Aquilae (Alshain) Beta and gamma (Tarazed) are the near-twin stars that flank Altair. Alshain is actually the fainter of the two. It shines at magnitude 3.7 compared to Tarazed's 2.7. It is just 49 light years away, while the giant Tarazed, which has a distinctly orange colour, is more than five times this distance.

NGC 6709 Aquila is crossed by a rich area of the Milky Way. NGC 6709 is an open cluster 3,000 light years away that appears through binoculars as a bright knot in the star clouds.

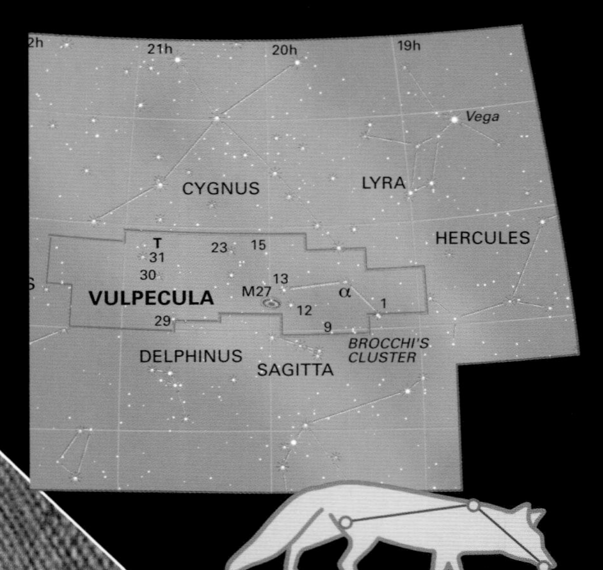

Vulpecula The Fox

Fully visible 90°N–61°S

This constellation is another of the additions made by Polish astronomer Johannes Hevelius in the late 17th century. It consists of a handful of faint stars with no obvious pattern, and is best located by looking to the west of the brighter constellation Pegasus.

FEATURES OF INTEREST

Alpha (α) Vulpeculae The constellation's brightest star has a modest magnitude of 4.4. It is a red giant about 250 light years away from Earth.

Brocchi's Cluster This small group of stars lies on Vulpecula's southern border. Its members hover at the limit of naked-eye visibility, and the cluster is an attractive sight in binoculars.

Dumbbell Nebula (M27) Vulpecula's most famous object, this is the brightest planetary nebula in the sky and among the easiest to spot. Appearing as a rounded patch, the nebula is one-quarter of the diameter of a full Moon, and about 1,000 light years from Earth. It is a rewarding target for small telescopes, which will reveal the twin lobes or hourglass shape for which it gets its popular name. The nebula usually appears grey-green.

Delphinus The Dolphin

Fully visible 90°N–69°S

Small and faint, but still distinctive, Delphinus lies just to the west of Pegasus. It represents either one of two dolphins from Greek myth – one sent by the sea god Poseidon to rescue the drowning lyre player Arion, or another sent to persuade the mermaid Amphitrite to be Poseidon's bride.

FEATURES OF INTEREST

Alpha (α) Delphini (Sualocin) This hot blue-white star is located 190 light years from Earth and shines at magnitude 3.8.

Beta (β) Delphini (Rotanev) Slightly brighter than alpha, Rotanev is a pure white star of magnitude 3.6 that lies 72 light years away from Earth. The names Sualocin (see above) and Rotanev are a reversed spelling of Nicolaus Venator, the Latinized name of an Italian astronomer called Niccolò Cacciatore working at Palermo Observatory in Sicily in around 1800, who ignored convention and mischievously named the stars after himself.

Gamma (γ) Delphini This attractive double star, about 125 light years away from Earth, consists of two yellow-white stars of magnitudes 4.3 and 5.1. They are easily separated with a small telescope. With Sualocin (α), Rotanev (β), and delta (δ), gamma forms the asterism known as "Job's Coffin", the name possibly being a reference to its boxlike or diamond shape.

Equuleus The Foal

Fully visible 90°N–77°S

Equuleus is the second smallest constellation in the sky, and its stars are relatively faint. It represents the head of a young horse and since ancient times has been seen as a companion to the nearby, larger horse-shaped constellation Pegasus, although there are no myths or legends associated with Equuleus. It is located by looking in the wedge of sky between Epsilon Pegasi, in the southwest corner of Pegasus, and the diamond-shaped constellation Delphinus.

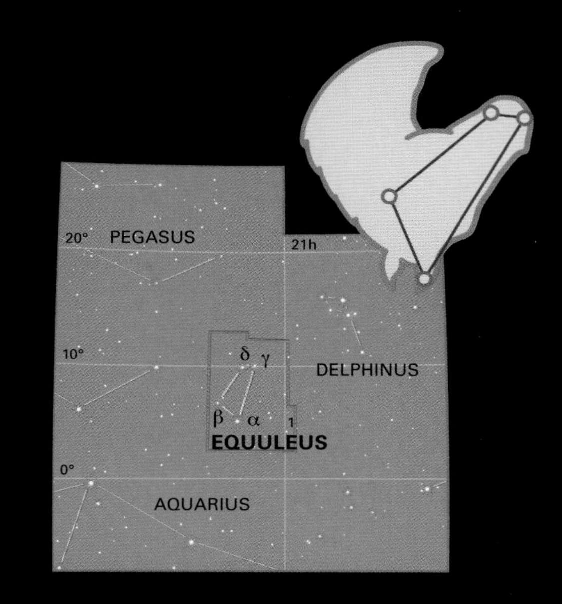

FEATURES OF INTEREST

Alpha (α) Equulei (Kitalpha) This yellow giant of magnitude 3.9 lies 190 light years from the Sun, and is 75 times more luminous than it.

Epsilon (ε) Equulei This triple star combines a chance alignment with a genuine binary system. A small telescope will reveal that the primary star, magnitude 5.4, has a magnitude-7.4 companion that just happens to lie in the same direction. The fainter star is actually the closer of the two, 125 light years away compared to 200 for the primary. Larger telescopes reveal that the primary is itself a double star.

Pegasus The Winged Horse

Fully visible 90°N–53°S

One of the largest constellations, Pegasus covers a rather empty area of sky. This constellation is very easy to find, because its four bright stars (one of which is shared with Andromeda) form the Great Square of Pegasus. It is one of the original 48 Greek constellations.

FEATURES OF INTEREST

Alpha (α) Pegasi (Markab) This is normally the brightest star in the constellation, blue-white and shining at magnitude 2.5. It is 140 light years away.

Beta (β) Pegasi (Scheat) This red giant shows an obvious colour difference from the other stars in the Great Square of Pegasus. It is 200 light years from Earth, and varies unpredictably – usually around magnitude 2.7, but sometimes outshining Markab, reaching magnitude 2.3, and occasionally becoming fainter than gamma at around magnitude 2.9.

M15 This bright globular cluster of magnitude 6.2 is easily spotted through binoculars. More than 30,000 light years away, it is one of the densest star clusters in the galaxy. It contains nine pulsars, the remains of ancient supernova explosions.

Aquarius The Water Carrier

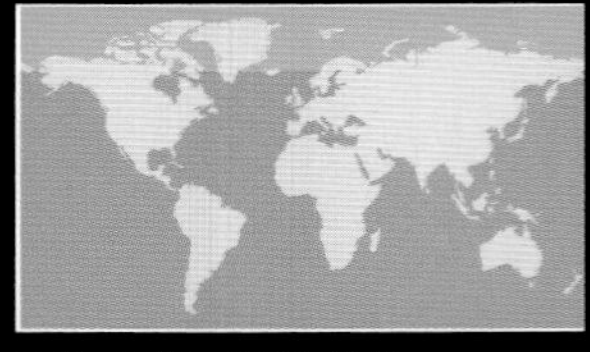

Fully visible 65°N–86°S

One of the oldest constellations, and a member of the zodiac, Aquarius has been seen as a youth (or, sometimes, an older man) pouring water from a jug since the second millennium BCE. The pattern is indistinct, but its brightest star and the Y-shaped pattern of the water jug are useful aids for locating it.

FEATURES OF INTEREST

Helix Nebula (NGC 7293) This is our nearest planetary nebula, at 300 light years away. One of the largest nebulae in apparent size, it appears the size of the full Moon. Its light, however, is diffused over a large area, so the nebula can be identified only in clear and dark skies. It is best seen through binoculars, which have a wide field of view.

Saturn Nebula (NGC 7009) Another planetary nebula, this is 3,000 light years away and shines at magnitude 8.0. It appears to be of a size similar to the disc of Saturn, when viewed with a small telescope.

M2 This is the brighter of Aquarius's two globular clusters, at magnitude 6.5. It is 37,000 light years away.

Pouring water
Aquarius is found between Capricornus and Pisces, near the celestial equator. The cascade of stars that represent the flow of water from Aquarius's jar is to the left of this image. The distinctive Water Jar is centre-top.

Pisces The Fish

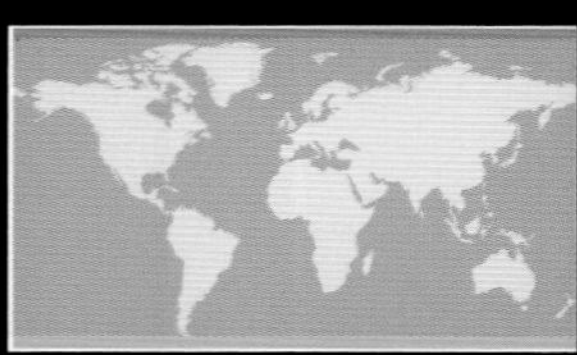

Fully visible 83°N–56°S

This zodiac constellation represents two mythical fish. It is best spotted by looking to one side of the Great Square of Pegasus. According to ancient Greek myths, the goddess Aphrodite and her son, Eros, transformed into fish and plunged into the River Euphrates to escape a fearsome monster called Typhon. In another version of the same story, two fish swam up and carried Aphrodite and Eros to safety on their backs.

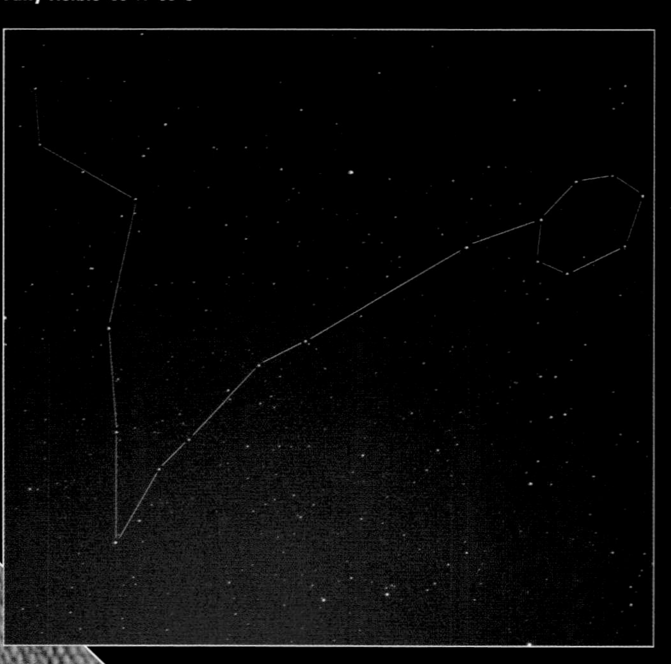

FEATURES OF INTEREST

Alpha (α) Piscium (Alrescha) Marking the point where the tails of Pisces's two fish join, Alrescha is a double star, though one whose components are too close in the sky to separate with small telescopes. Its white stars are 140 light years away, and have individual magnitudes of 4.2 and 5.2, giving them a combined brightness of magnitude 3.8.

Eta (η) Piscium The second brightest star in Pisces, eta is a yellow supergiant of magnitude 3.6 – brighter than either of alpha's components alone, and more than twice their distance at 300 light years.

M74 This spiral galaxy, 25 million light years away, is face-on to Earth. Its light is so spread out that it is quite a challenging target for small telescopes.

The Circlet
The most distinctive feature of Pisces is the ring of seven stars, seen on the top right. Known as the Circlet, it lies to the south of the Great Square of Pegasus and marks the body of one of the fish.

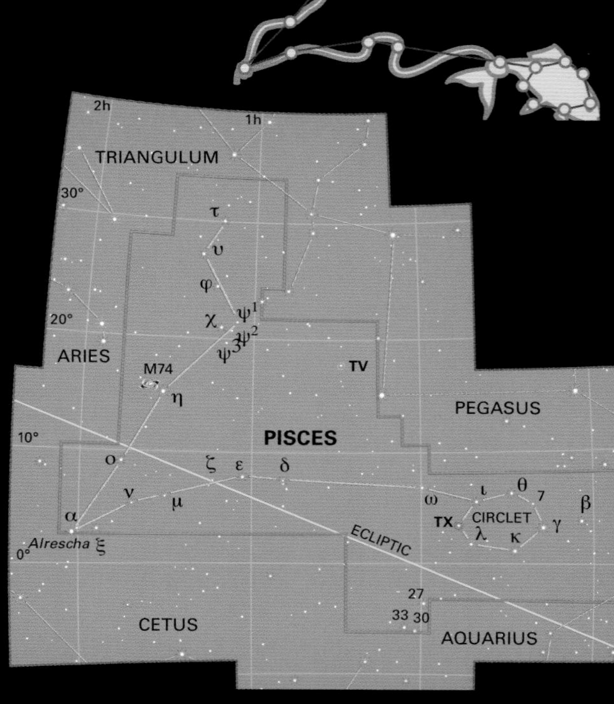

Cetus The Sea Monster or Whale

Fully visible 65°N–79°S

Usually seen as the figure of a whale, Cetus doubles as a sea monster for one of the sky's great legends – the story of Perseus and Andromeda. Cetus is large but relatively faint, best found by looking next to Taurus. Identification is not helped by the fact that it can vary in appearance depending on the brightness of its most famous star, Mira.

Lurching monster
Cetus is one of the original 48 Greek constellations listed by Ptolemy in his *Almagest*. It can be seen in the equatorial region of the sky, lying south of the constellations Pisces and Aries.

FEATURES OF INTEREST

Omicron (ο) Ceti (Mira) With a name derived from the Latin for "wonderful", Mira is among most prominent variable stars in the sky, and was recognized in 1596. It is distinctively red and varies between magnitude 10 and 2 over a cycle of 332 days. Mira is an unstable red giant, and it varies in brightness as it fluctuates in size. Depending on how much it has swollen or contracted within its cycle, Mira can be either a naked-eye star or one that is visible only with a telescope.

Tau (τ) Ceti One of the closest Sun-like stars to Earth, tau is just 11.9 light years away. Technically, it is a "yellow sub-dwarf", and if it has planets in orbit around it, they would be prime candidates for extraterrestrial life.

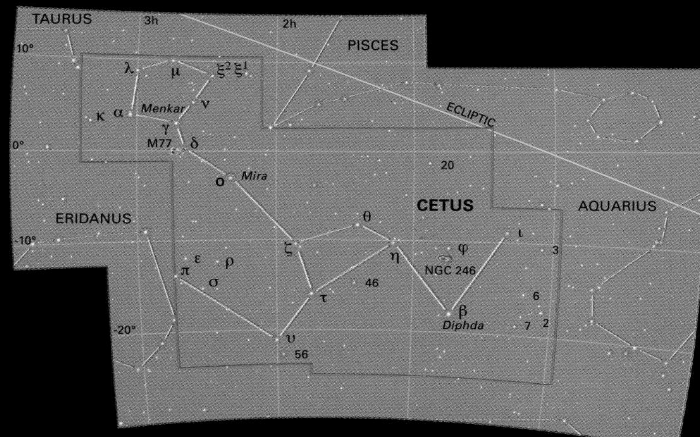

Orion Orion

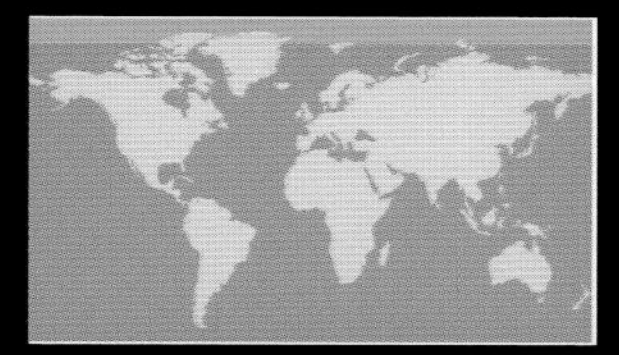

Fully visible 90°N–57°S

This prominent constellation, containing two of the brightest stars and the clearest emission nebula in the sky, represents a hunter from Greek myth. Orion stands in a celestial tableau, facing the charging bull Taurus, and followed by his faithful hounds, Canis Major and Canis Minor.

FEATURES OF INTEREST

Alpha (α) Orionis (Betelgeuse) Some 430 light years from Earth, this is one of the brightest red giants in the sky. But its magnitude varies unpredictably, between magnitude 0.0 and 1.3. It usually shines around 0.1, and is so large that astronomers have been able to map its surface.

Beta (β) Orionis (Rigel) Apart from the rare times when Betelgeuse is at its maximum magnitude, Rigel is Orion's brightest star. It is a brilliant blue-white supergiant of magnitude 0.1, and marks one of Orion's feet. Rigel lies about 770 light years away.

Great Orion Nebula (M42) Forming a "sword" below the three stars of Orion's belt, M42 is a huge star-forming region about 1,500 light years away. The nebula and its surrounding stars are visible to the naked eye, and a beautiful sight through binoculars or a small telescope. At its heart lies a small cluster of four recently formed stars known as the Trapezium.

Bright hunter
Orion is one of the most magnificent and easily recognizable constellations. A line of three stars makes up the hunter's belt, while an area of star clusters and nebulae forms his sword.

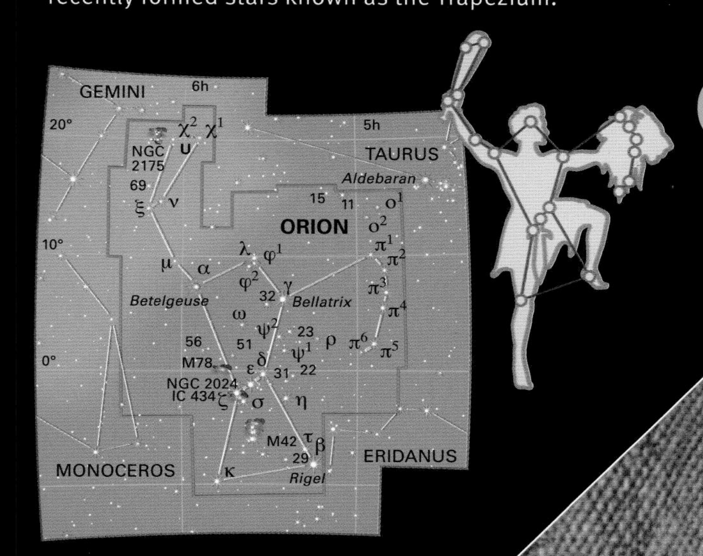

Canis Major The Greater Dog

Fully visible 56°N–90°S

Following obediently behind Orion the hunter on his journey across the sky, Canis Major is host to Sirius, the Dog Star, the brightest star in the sky. But since it lies across the Milky Way, it also contains several star clusters and other features of the deep sky.

FEATURES OF INTEREST

Alpha (α) Canis Majoris (Sirius) With a magnitude of -1.4, the famous "Dog Star" is so bright that only some planets ever outshine it. At 23 times more luminous than the Sun, Sirius is fairly average for a white star of its type, but it also happens to lie in our stellar neighbourhood, just 8.6 light years away. It is a binary system: the primary star is orbited by Sirius B – a faint white dwarf that would be more easily seen were it not for Sirius's own brilliance.

Beta (β) Canis Majoris (Mirzam) Although at magnitude 2.0, Mirzam is far outshone by Sirius, it is a much more luminous star in reality – a blue giant 500 light years from Earth.

M41 This open cluster of stars is visible to the naked eye as a hazy patch the size of the full Moon, although it is 2,300 light years away. Binoculars distinguish its brightest stars, while telescopes show chains of stars radiating from its centre.

Canis Minor The Lesser Dog

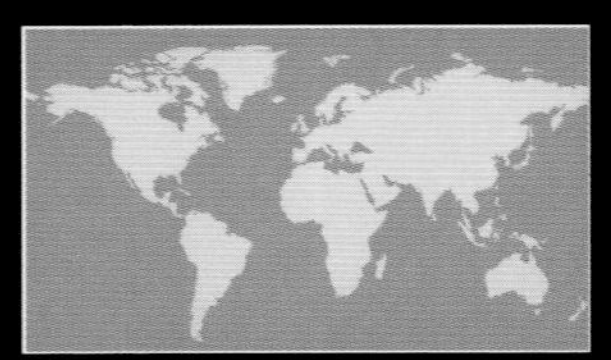

Fully visible 89°N–77°S

This small constellation is easy to spot because of its bright star Procyon. It is a Greek constellation and represents the smaller of Orion's two dogs. It forms an obvious triangle in the sky, with Sirius in Canis Major and Betelgeuse in Orion. Its border lies almost on the celestial equator.

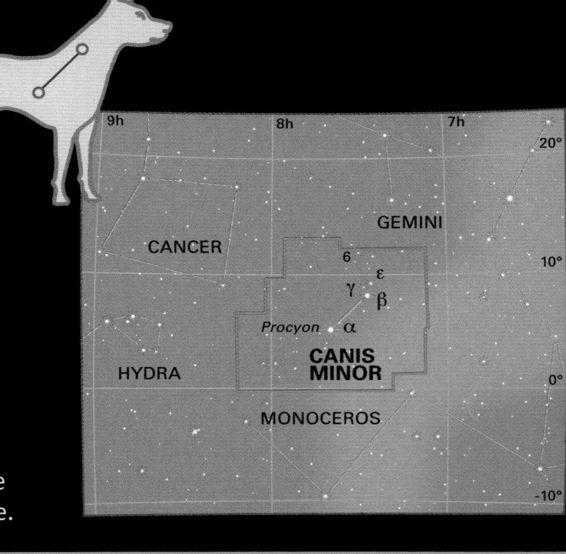

FEATURES OF INTEREST

Alpha (α) Canis Minoris (Procyon) The name of this star means "before the dog" in Greek, because from Mediterranean latitudes, it always rises shortly before the more brilliant Sirius. Shining at magnitude 0.34, it is still one of the most prominent stars in the sky, and offers a useful comparison with its brighter cousin, since both stars lie around the same distance from Earth. In reality, Procyon is seven times more luminous than the Sun, compared to Sirius's 23 times. Like Sirius, it forms a binary system with a white dwarf companion, named Procyon B.

Beta (β) Canis Minoris (Gomeisa) This magnitude-2.9 blue-white star is far more distant than Procyon, at about 150 light years away, and is much more radiant. Its name derives from the Arabic for "the little bleary-eyed one" which referred to the weeping sister of Sirius, whom he left behind to flee for his life.

Monoceros The Unicorn

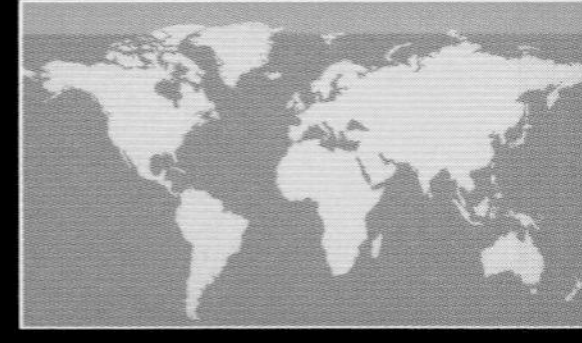

Fully visible 78°N–78°S

The W-shape of Monoceros is hard to pick out, but it can be located with reference to Orion and Canis Major. The constellation sits on the celestial equator in the middle of a triangle formed by Betelgeuse (in Orion), Procyon (in Canis Minor), and Sirius (in Canis Major). Added to the list of constellations by Dutch theologian Petrus Plancius in 1613, it depicts a unicorn.

FEATURES OF INTEREST

Alpha (α) Monocerotis This is the brightest of the constellation's stars. It is an orange giant, about 175 light years from Earth and shining at magnitude 3.9.

Beta (β) Monocerotis The constellation's stellar highlight, beta is a beautiful triple star, separated in a small telescope to reveal a chain of three fifth-magnitude, blue-white stars.

M50 This is one of several open star clusters populating a rich band of the Milky Way as it passes through Monoceros. Small telescopes should reveal its individual stars.

NGC 2244 This star cluster lies at the heart of a glowing gas cloud called the Rosette Nebula – an outlying part of a huge star-forming complex centred on Orion. The Rosette itself (NGC 2237) is diffuse, but can be seen with good binoculars on a dark night.

Hydra The Water Snake

The largest constellation in the sky is a hard-to-trace chain of mostly average-brightness stars. The head of this lengthy snake is a roughly triangular group of stars south of Cancer. The brightest star, Alphard ("solitary one"), marks its heart.

Fully visible 54°N–83°S

FEATURES OF INTEREST

M48 This open cluster of stars lies close to Hydra's border with the richer starfields of Monoceros. It contains about 80 stars, and is just visible to the naked eye in dark skies.

M83 This face-on spiral galaxy, 15 million light years away, has a bright central nucleus that can be spotted with ease through a small telescope.

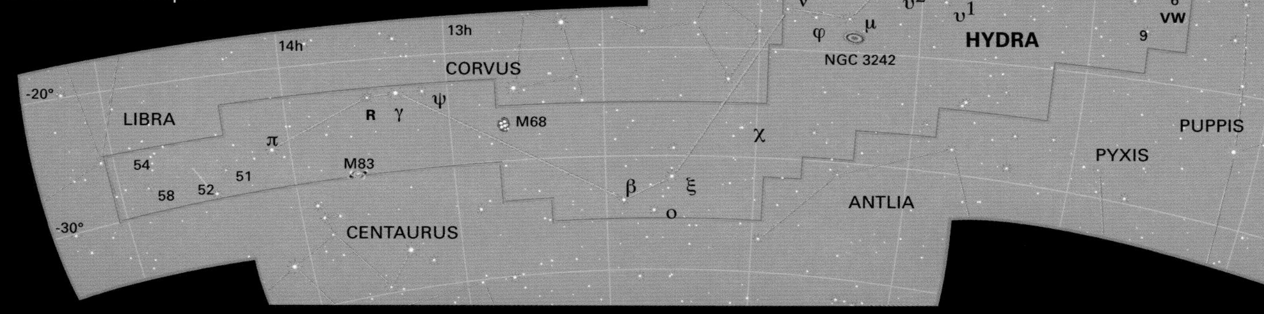

Long serpent
The Hydra's head, at the right in this photograph, lies south of Cancer (and, in this view, the disc of Jupiter), while the tip of its tail lies far to the left.

Antlia The Air Pump

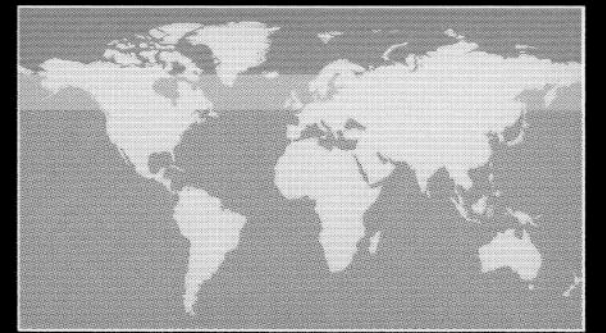

Fully visible 49°N–90°S

French astronomer Nicolas de Lacaille introduced this constellation for his 1756 map of the southern skies. It honours the vacuum pump invented by French scientist Denis Papin and British physicist Robert Boyle. It is best found by looking to the northeast of the Milky Way as it passes through Puppis.

FEATURES OF INTEREST

Alpha (α) Antliae This orange giant is 500 times more luminous than the Sun, but at a distance of 365 light years, it shines at a weak magnitude of 4.3.

Theta (θ) Antliae At magnitude 4.8, theta is the constellation's second brightest star, though it is actually a double, consisting of white and yellow stars of magnitudes 5.6 and 5.7 respectively, 385 light years away. Unfortunately, small telescopes are not powerful enough to separate them.

Eight-burst Nebula (NGC 3132) Sometimes also referred to as the Southern Ring Nebula, this nebula straddles the boundary of Antlia and Vela, at a distance of about 2,000 light years from Earth. It is a planetary nebula, which often forms when a Sun-like star becomes a red giant and throws off its outer layers. NGC 3132 shines at magnitude 8 and so is a good target for small telescopes.

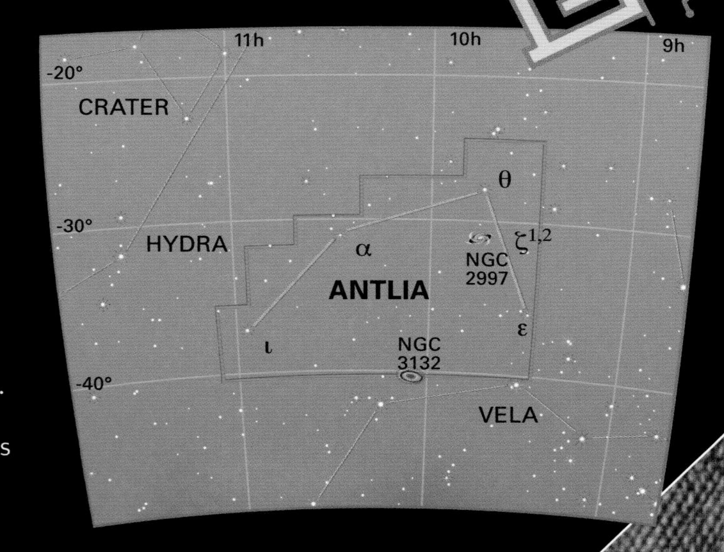

North of Vela
Antila is an inconspicuous grouping in the southern hemisphere and consists of a handful of stars to be found between Vela and Hydra.

Sextans The Sextant

Fully visible 78°N–83°S

This constellation is named after a navigational and scientific instrument used for taking star positions in the days before telescopes. The pattern is easy to find, since it lies to the south of Leo's bright star, Regulus. It was added to the sky by Polish astronomer Johannes Hevelius in 1687.

FEATURES OF INTEREST

Alpha (α) Sextantis This blue-white giant star is some 340 light years from Earth. Due to this distance, alpha shines in Earth's skies at a relatively weak magnitude 4.5.

Beta (β) Sextantis Another blue-white giant, beta is more luminous than alpha, but only reaches magnitude 5.1 in our skies because it lies 520 light years away.

Spindle Galaxy (NGC 3115) One of the closest large galaxies, NGC 3115 lies some 14 million light years away (many dwarf elliptical galaxies lie closer but are generally too faint for amateur observers). Usually classified as lenticular (lens-shaped) galaxy, NGC 3115's huge, bulging disc of stars can appear elliptical, because it is viewed edge-on from Earth. The combined light of its stars reaches magnitude 8.5, making it just visible in binoculars, though a telescope of small to moderate aperture is needed for a proper view. It is popularly named the Spindle Galaxy because of its highly elongated shape.

Crater The Cup

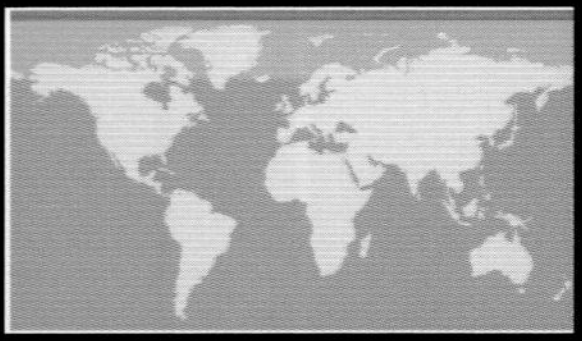

Fully visible 65°N–90°S

This constellation is faint, but still relatively easy to locate due to its distinctive "bow-tie" shape. It represents the drinking cup of the Greek god Apollo, and is linked in mythology with its adjacent constellations. Supposedly Apollo sent Corvus, the crow, to fill his cup from a well, but the bird was distracted by a fig tree, and returned with an empty cup, saying that Hydra, the water-snake, had blocked the well. An angry Apollo saw through the crow's deception, and threw snake, cup, and crow into the sky, to be preserved among the stars.

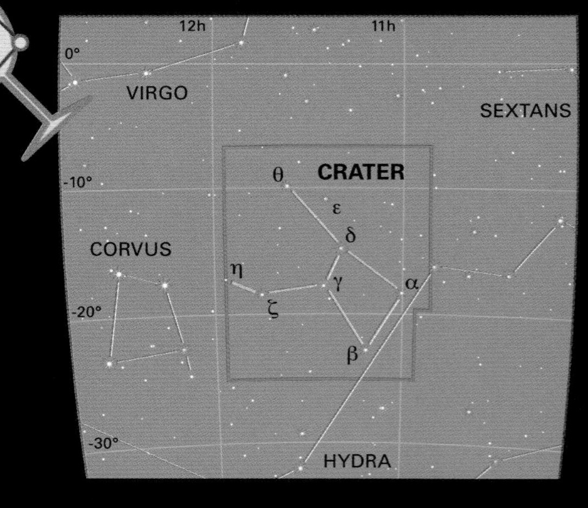

FEATURES OF INTEREST

Delta (δ) Crateris Crater's brightest star has ended up with the designation delta through historical accident. It is a magnitude-3.6 orange giant, 62 light years from Earth.

Alpha (α) Crateris Significantly fainter than gamma at magnitude 4.1, alpha is a yellow giant star, about 175 light years away.

Gamma (γ) Crateris This white star, 75 light years away and of magnitude 4.1, has a faint binary companion that can be seen in small telescopes.

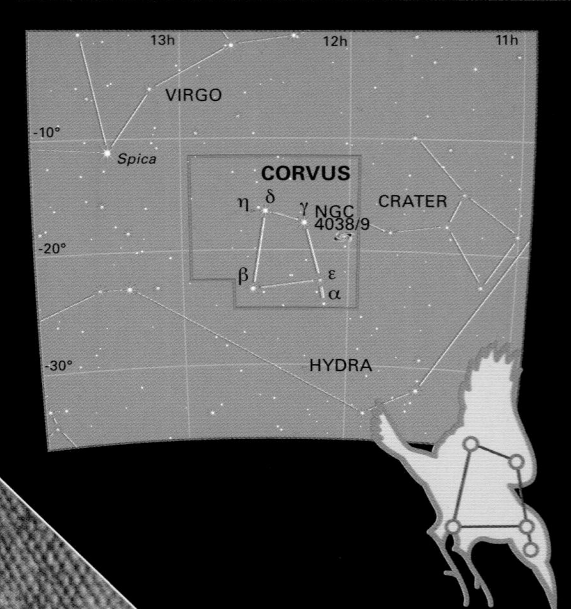

Corvus The Crow

Fully visible 65°N–90°S

The roughly rectangular shape of Corvus is defined by its four brightest stars and represents a crow, the servant of the god Apollo in a story linking it with Crater and Hydra. Its pattern is indistinct, and it is best found by looking to the southwest of the bright star Spica in Virgo.

FEATURES OF INTEREST

Gamma (γ) Corvi (Gienah) Corvus's brightest star, gamma is a blue-white star of magnitude 2.6, lying at a distance of 220 light years. It shares a common name with Epsilon (ε) Cygni.

Delta (δ) Corvi This double star, 115 light years distant, is a good target for small telescopes. The bright blue-white primary is orbited by a deeper blue or even purple star of magnitude 9.2.

Alpha (α) Corvi (Alchiba) Despite its Bayer letter designation, usually given to the brightest star, alpha is outshone by Gamma (γ), Beta (β), and Delta (δ) Corvi. It is a white star 52 light years away, shining at magnitude 4.0.

Antennae (NGC 4038 and 4039) These faint galaxies, barely visible through small telescopes, are a pair of colliding spiral galaxies. Their name describes the long tendrils of stars, gas, and dust ripped from the galaxies during their encounter.

Centaurus The Centaur

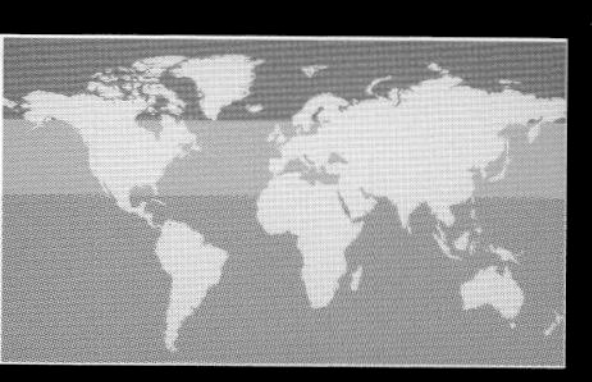

Fully visible 25°N–90°S

Centaurus extends into the Milky Way and contains several deep-sky objects as well as the closest star to Earth. This constellation represents a mythical centaur called Chiron.

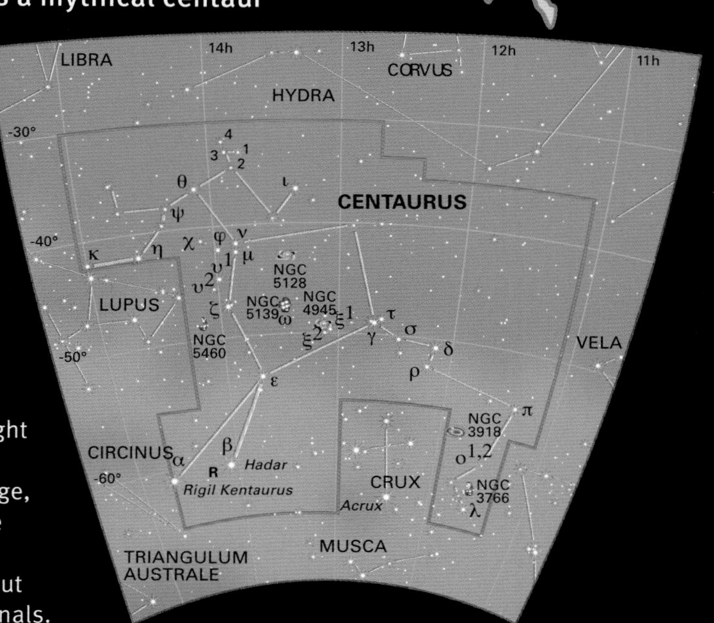

FEATURES OF INTEREST

Alpha (α) Centauri (Rigil Kentaurus) The sky's third-brightest star at magnitude -0.3, this system is our next-door neighbour, 4.3 light years away. This brilliant star is in fact a yellow star with a red companion of magnitudes 0 and 1.3 respectively. Another companion, the 11th-magnitude red dwarf Proxima Centauri, can be spotted only through a good telescope.

Omega (Ω) Centauri (NGC 5139) Despite its stellar name, omega is the sky's brightest globular cluster – a tight ball of several million stars, 17,000 light years from Earth and shining at magnitude 3.7. To the naked eye, it is a large, hazy star, and a small telescope is required to resolve the brightest individual members of this globular cluster.

NGC 5128 This bright galaxy is an active elliptical, about 15 million light years away, that gives out strong radio signals.

Celestial centaur
The brilliant stellar pairing of Alpha (α) and Beta (β) Centauri guides the eye to Centaurus, the celestial centaur. The familiar pattern of Crux, the Southern Cross, lies beneath the centaur's body.

Bestial offering
Here, Lupus is partly surrounded by the stars of Centaurus. In Greek and Roman myths, Lupus represented a wild animal that had been speared on a long pole by Centaurus. The identification of Lupus as a wolf seems to have become common during Renaissance.

Lupus The Wolf

Fully visible 34°N–90°S

Lying in the southern Milky Way and cursed with a complex jumble of stars, Lupus can be hard to spot, despite being relatively bright. However, it lies between the more recognizable Scorpius and Centaurus and contains many interesting objects.

FEATURES OF INTEREST

Alpha (α) and Beta (β) Lupi Lupus's two brightest stars are almost identical and close neighbours in space. Both are blue giants about 650 light years away, but alpha is slightly closer, giving it a magnitude of 2.3 compared to beta's 2.7.

Mu (μ) Lupi Many of Lupus's stars are double, but mu is one of the easiest to split. Small telescopes will easily show that the primary, a blue-white star of magnitude 4.3, has a magnitude-7 companion. Larger telescopes will reveal that the primary is itself a double, composed of twin stars of magnitude 5.1.

NGC 5822 This large open cluster of stars, about 2,600 light years away, has an overall magnitude of 7.0.

NGC 5986 This globular cluster also shines at magnitude 7 – its stars are much more distant than those of NGC 5822, at around 45,000 light years, but they are also far more numerous.

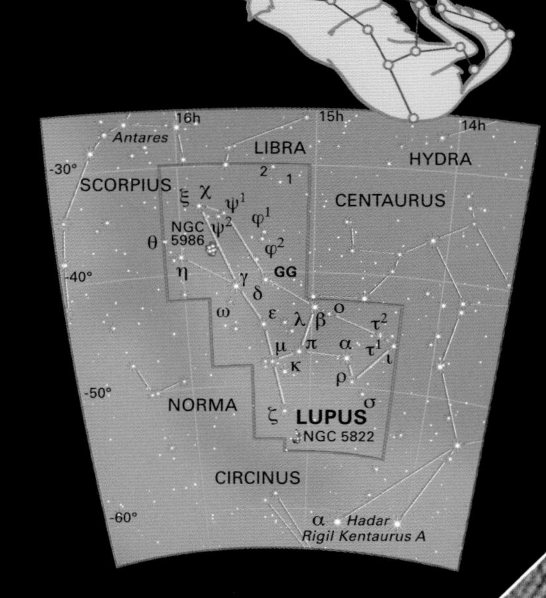

Mounted bowman
The prominent constellation of Sagittarius is found between Scorpius and Capricornus, in the southern celestial hemisphere. In Greek mythology, Sagittarius was also identified with the satyr Crotus, son of Pan.

Sagittarius The Archer

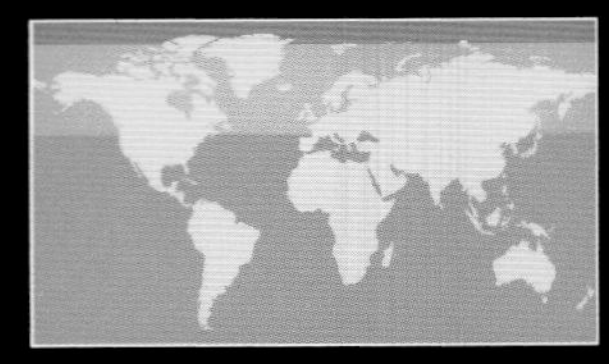

Fully visible 44°N–90°S

Usually represented as a centaur (half man, half horse) armed with a bow, Sagittarius lies in the richest area of the Milky Way. Its central stars are best spotted by looking for a teapot-shaped pattern in the sky.

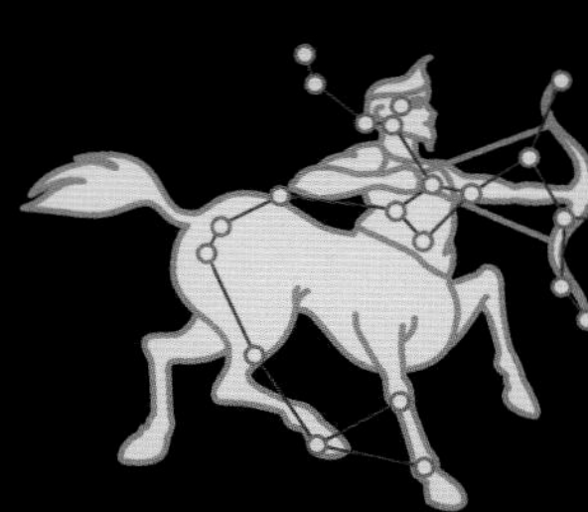

FEATURES OF INTEREST

Sigma (σ) Sagitarii (Nunki) This blue-white star is the constellation's brightest, with a magnitude of 2.0.

Beta (β) Sagittarii (Arkab) This apparent double star, consisting of two stars at about magnitude 4.0, can be split with the naked eye – but its stars are a chance alignment – in reality they are 140 and 380 light years away.

Lagoon Nebula (M8) Sagittarius is rich in deep-sky objects, as it lies in the direction of the galactic centre. Highlights are a chain of star-forming nebulae, including the Omega Nebula (M17). Largest and brightest of all is M8, visible to the naked eye as a lighter patch of sky and easily identified in binoculars.

M22 Visible to the naked eye and a fine sight in binoculars, this is the brightest of several globular clusters on the northern edge of the constellation.

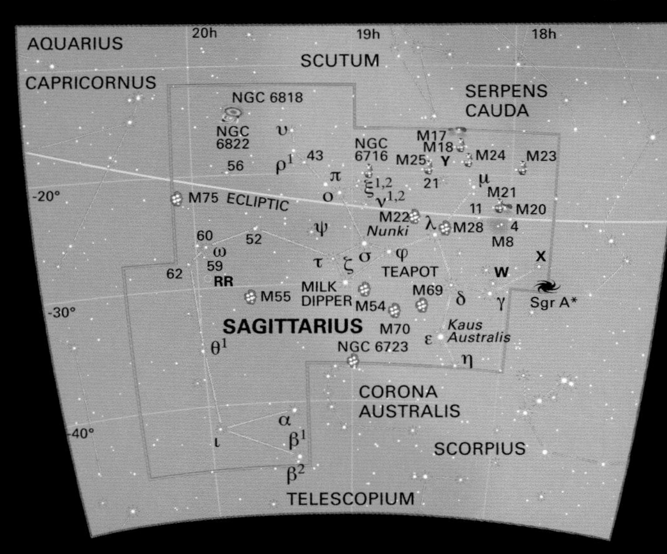

Scorpius The Scorpion

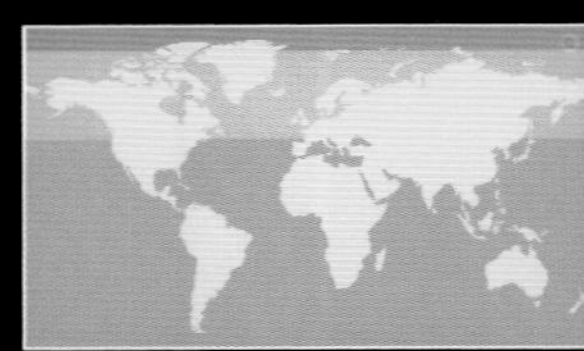

Fully visible 44°N–90°S

This ancient zodiac constellation contains many dense Milky Way star clouds. Scorpius represents a scorpion that killed the hunter Orion in Greek myth – hence its location on the opposite side of the sky.

FEATURES OF INTEREST

Alpha (α) Scorpii (Antares) With a name that means "rival to Mars", this brilliant star varies in brightness between magnitudes 0.9 and 1.8 over a six-year period. Near-identical flanking stars Sigma and Tau Scorpii make it even easier to identify. Antares is a red supergiant, hundreds of times larger than the Sun and lying 600 light years from Earth.

M6 This fine open star cluster is visible to the naked eye as a "knot" in the Milky Way, just above the scorpion's tail. Binoculars or a small telescope will reveal dozens of individual stars. M6 lies 2,000 light years away, while its chance neighbour M7 is just 800 light years from Earth, and so is rather brighter.

M4 This globular cluster, 7,000 light years away and in orbit around the Milky Way, shines at magnitude 7.4, making it a good target for binoculars or a telescope.

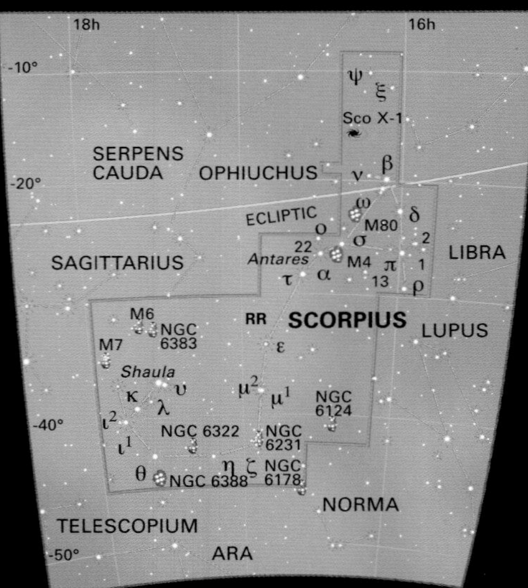

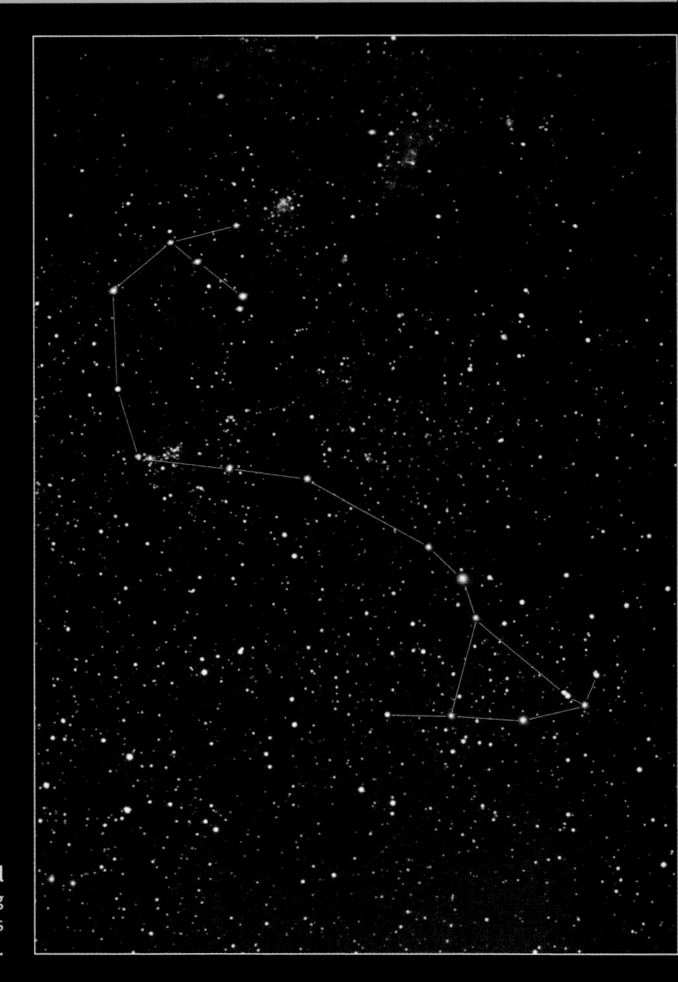

Sting in the tail
This view of Scorpius shows the scorpion raising its curving tail as though to strike. Its heart is marked by the red star Antares.

Capricornus The Sea Goat

Less recognizable than its southeastern neighbour Sagittarius, this constellation represents one of the stranger creatures in the sky, half fish and half goat. In ancient Greek legend, Capricornus was supposed to be the goat-headed god Pan, depicted in the sky escaping danger by transmuting into a fish.

Fully visible 62°N–90°S

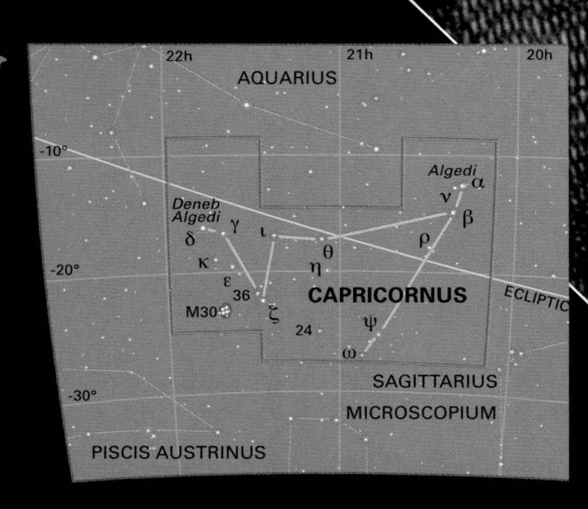

FEATURES OF INTEREST

Alpha (α) Capricorni (Algedi) This is an impressive multiple star, though not all its elements are truly related. The brightest stars, easily separated with binoculars or good eyesight, are a yellow supergiant (α^1) and an orange giant (α^2), 690 and 109 light years away and with magnitudes of 4.2 and 3.6. Small telescopes show that the yellow supergiant is a double star in its own right; larger ones will reveal that the orange giant is in fact a triple star.

Beta (β) Capricorni (Dabih) This magnitude-3.3 yellow giant reveals a faint companion through binoculars. Beta is in fact a complex multiple containing at least five, and perhaps eight, stars in orbit around each other, 330 light years away.

M30 This globular cluster, of magnitude 7.5, is 27,000 light years away and visible through binoculars. Chains of stars extend like fingers from the northern side of the cluster.

Microscopium The Microscope

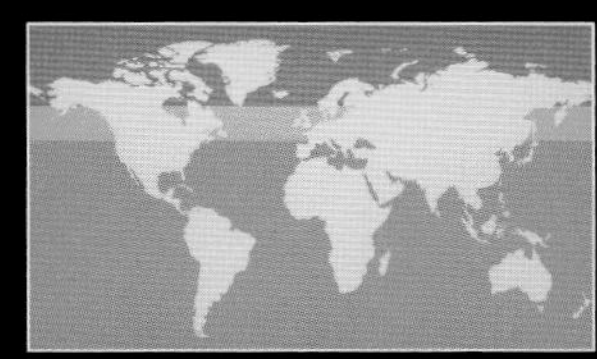

Fully visible 45°N–90°S

This constellation is one of several small, faint groups added to the sky by French astronomer Nicolas Louis de Lacaille during the 1750s, most of them named after scientific instruments. Microscopium is best found by looking between the brighter stars of Sagittarius and Fomalhaut (in Piscis Austrinus).

FEATURES OF INTEREST

Alpha (α) Microscopii This double star lies 250 light years from Earth. The primary is a yellow giant of magnitude 5.0, while its companion is far fainter at magnitude 10, and visible only through telescopes of moderate aperture.

Gamma (γ) Microscopii Slightly brighter than alpha at magnitude 4.7, gamma is a yellow giant 245 light years away. To its southeast is the star epsilon (ε) Microscopii, while alpha (α) Microscopii lies at its northwest.

Theta (θ) Microscopii This is the brightest of several variable stars in the constellation, but its variations are hard to see, since it varies by only about 0.1 magnitudes from its average of 4.8 in a cycle lasting 2 days.

U Microscopii This distant red giant is a more obvious variable – it pulsates in the same way as the famous Mira in Cetus, varying between magnitudes 7.0 and 14.4 in 334 days. S Microscopii is similar but has a shorter 209-day cycle.

Piscis Austrinus The Southern Fish

Fully visible 53°N–90°S

This small ring of generally faint stars is made more obvious by the presence of Fomalhaut, one of the brightest stars in the sky. Piscis Austrinus was invented in ancient times and is one of the most southerly members of a list of 48 constellations drawn up the Greek astronomer Ptolemy.

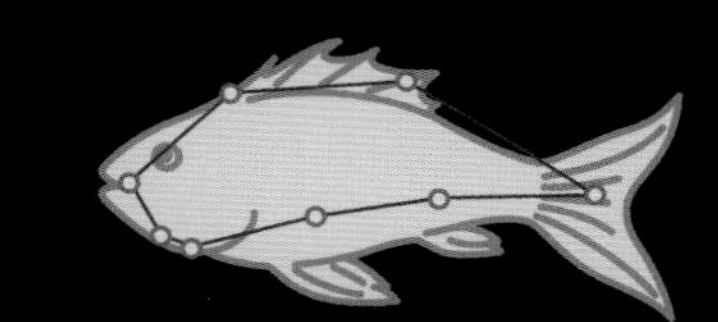

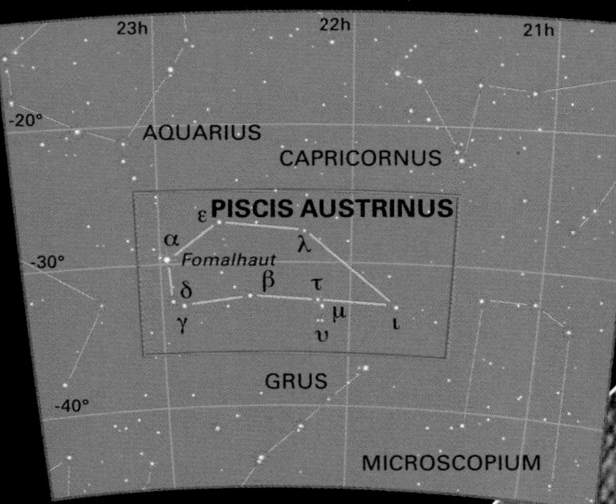

FEATURES OF INTEREST

Alpha (α) Piscis Austrini (Fomalhaut) At magnitude 1.2, Fomalhaut is the 18th brightest star in the sky. It is also relatively nearby, just 22 light years away, which means that in reality it is some 16 times more luminous than the Sun. Fomalhaut was the first star found to have a disc around it – in this case, a ring of cold, icy material, extending over about twice the diameter of our own solar system. It seems that Fomalhaut, which is relatively young, is still in the process of forming its own planets.

Beta (β) Piscis Austrini This double star, 135 light years away, consists of a magnitude-4.3 primary with a widely spaced magnitude-7.7 star easily seen through a small telescope.

Gamma (γ) Piscis Austrini A more challenging double star, gamma's primary, of magnitude 4.5, has a closer companion of magnitude 8.0. The system lies 325 light years from Earth.

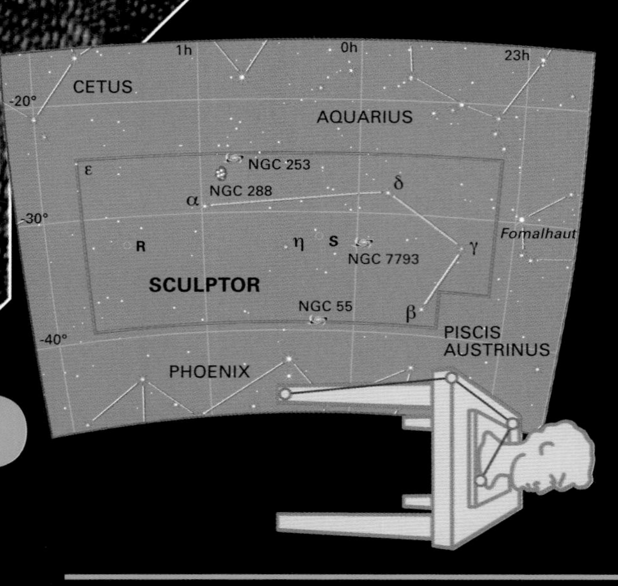

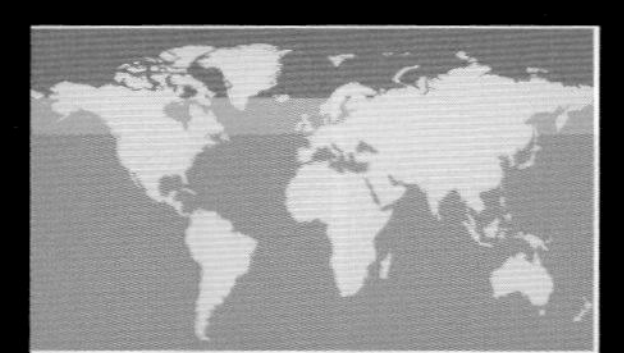

Sculptor The Sculptor's Studio

Fully visible 50°N–90°S

Representing a sculptor's workshop, this is one of the more bizarre constellations added to the sky by French astronomer Nicolas Louis de Lacaille.Its stars are faint and their pattern uninspiring, but the constellation is redeemed by the presence of several interesting nearby galaxies.

FEATURES OF INTEREST

Alpha (α) Sculptoris The brightest star in the constellation is a blue-white giant, 590 light years away and shining at magnitude 4.3.

NGC 55 This 8th-magnitude galaxy is only 6 million light years away, in the galaxy cluster just beyond the edge of our own Local Group. A spiral galaxy, it is mottled with dust clouds and areas of star formation. Though just visible through binoculars, it an easier target for small telescopes.

NGC 253 The largest and brightest member of the so-called Sculptor Group, this spiral galaxy is about 9 million light years away, at the heart of its galaxy cluster. It shines at around magnitude 7.5 and appears through binoculars as a fuzzy oval of light with the same diameter as the full Moon. The central region of the galaxy appears as a bright, starlike point – an indication that its heart is unusually active. Like NGC 55, it appears edge-on but it is brighter and easier to view with binoculars.

Fornax The Furnace

Fully visible 50°N–90°S

Another of the constellations introduced by Nicolas Louis de Lacaille during his observations from the Cape of Good Hope in the early 1750s, Fornax was originally Fornax Chemica, the Chemist's Furnace. It is mostly enclosed between the larger constellations of Eridanus and Cetus.

FEATURES OF INTEREST

Alpha (α) Fornacis The constellation's brightest star is double, easily split in small telescopes to reveal a magnitude-3.9 yellow star with an orange companion that has a magnitude of 6.9.

NGC 1097 Sixty million light years away, and shining at magnitude 10.3, this is one of the sky's brightest barred spiral galaxies. A small telescope will show its bright central nucleus, but a larger instrument is needed to show the barred structure and a dark bar of dust through the centre. Astronomers now believe that our own Milky Way galaxy is also a barred spiral.

NGC 1316 This unusual galaxy is associated with a strong radio source called Fornax A. It appears to be an elliptical galaxy that has recently absorbed another one. Dust and gas falling into the larger galaxy have awakened its central black hole, giving it an unusually bright and active core.

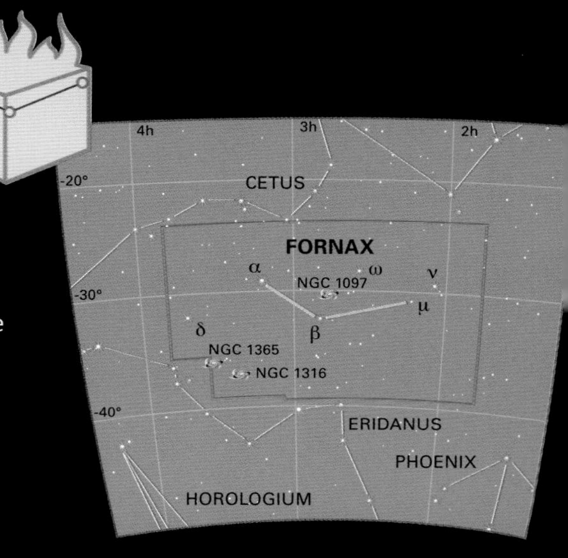

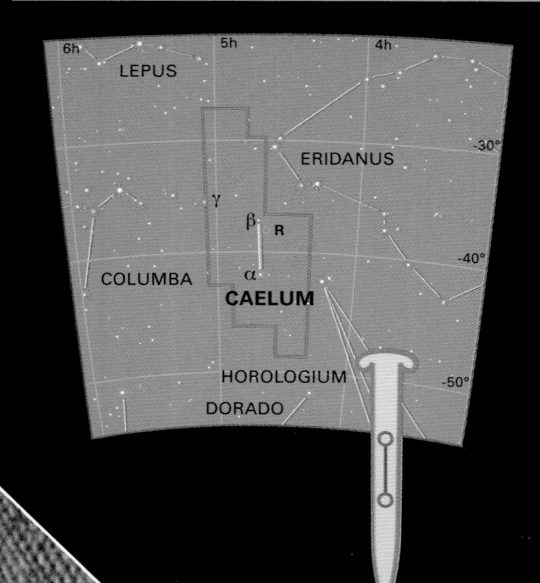

Caelum Caelum

Fully visible 41°N–90°S

Caelum is another of French astronomer Nicolas Louis de Lacaille's additions to the southern sky and has an unenviable reputation as one of the least impressive constellations in the heavens. Its pattern of two faint stars is supposed to represent an 18th-century engraver's chisel.

FEATURES OF INTEREST

Alpha (α) Caeli The brightest star in Caelum is a paltry magnitude 4.5. It is a white star 62 light years from Earth.

Beta (β) Caeli Like alpha, this is a white star of average luminosity. It shines at magnitude 5.1, and is a close neighbour in space of alpha, lying some 65 light years away.

Gamma (γ) Caeli Sitting on the constellation's western boundary, gamma is a magnitude-4.6 orange giant, some 280 light years away. A small telescope will reveal that it is a binary star, with a faint companion of magnitude 8.1.

R Caeli Lying just to the south of Beta Caeli, R is a variable star. Similar in type to Mira in Cetus, it is a slowly pulsating red giant with a long period of around 400 days. At its peak, it is of magnitude 6.7 and easily spotted in binoculars, but at its dimmest, around 13.7, it sinks beyond the range of small telescopes.

Eridanus The River

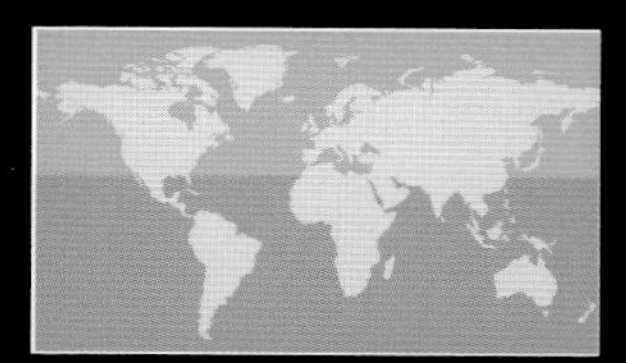

Fully visible 32°N–90°S

The elongated form of Eridanus, the celestial river, leads from the foot of Orion deep into southern skies.

FEATURES OF INTEREST

Alpha (α) Eridani (Achernar) With a name meaning "river's end" in Arabic, the constellation's one truly bright star marks its southern tip. Achernar is a blue-white giant of magnitude 0.5, some 95 light years from Earth.

Epsilon (ε) Eridani On the constellation's northern stretch, epsilon is one of the closest Sun-like stars to Earth. It lies 10.5 light years away and shines at magnitude 3.7. It is a little fainter and cooler than our Sun, and a planet-forming disc of gas and dust surrounds it.

Omicron 2 (o2) Eridani This triple-star system is 16 light years from Earth and hosts the most easily seen white dwarf star in the sky. The primary is a magnitude-4.4 red dwarf, but its companion is a white dwarf of magnitude 9.5, which itself has a fainter red dwarf partner.

Celestial river
Eridanus has its source next to Rigel (in Orion) and flows south to Achernar. It is fully visible to almost all of the southern hemisphere and half of the Northern.

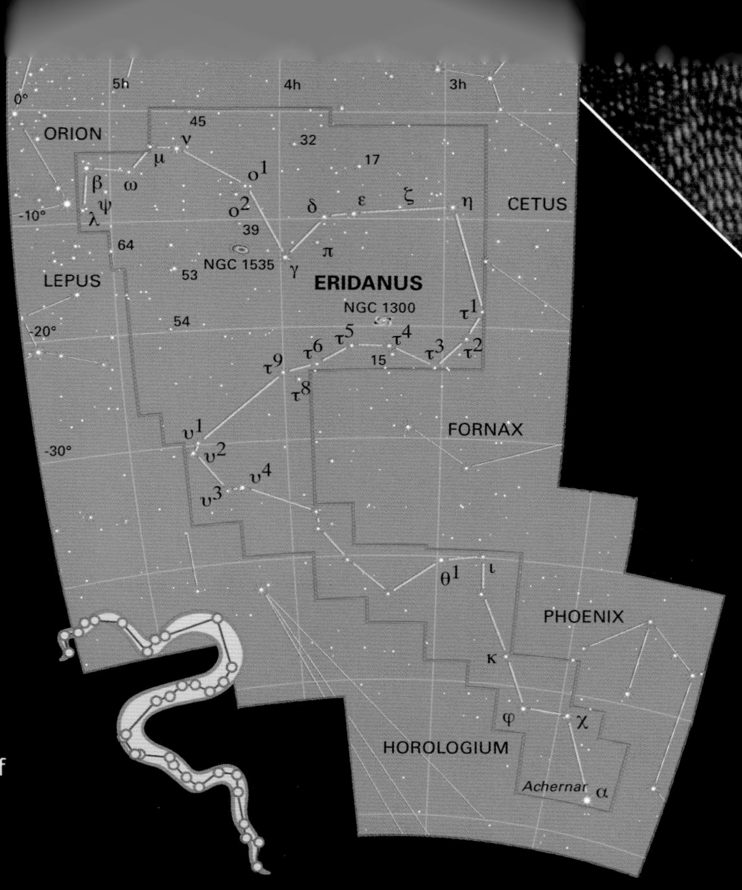

Lepus The Hare

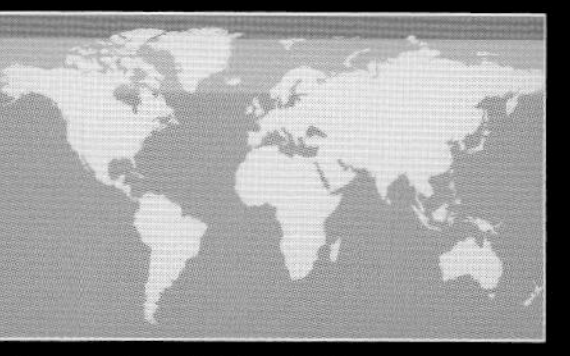

Fully visible 62°N–90°S

Lying just to the south of Orion, Lepus represents a hare at the feet of the celestial hunter, hotly pursued by the dog Canis Major. It was known to the ancient Greeks and is relatively easy to spot among its sparkling neighbours, since its brighter stars form a distinctive bow-tie shape.

FEATURES OF INTEREST

Alpha (α) Leporis At magnitude 2.6, this star seems to be of average brightness, but it lies a distant ,300 light years away and is actually one of the most luminous stars visible from Earth.

Gamma (γ) Leporis Gamma is a double star, with a yellow primary of magnitude 3.9 and a magnitude-6.2 orange companion. Both stars are at a similar distance from Earth, about 30 light years away.

R Leporis (Hind's Crimson Star) This object is a pulsating red giant variable, noted for its deep red colour. It ranges in magnitude between 5.5 and 12.0 over a 430-day cycle.

NGC 2017 This beautiful multiple star consists of eight colourful stars in all, five of which lie between magnitudes 6 and 10, and so are visible in binoculars.

M79 This globular cluster, 42,000 light years away and shining at magnitude 7.7, may have originated in a small dwarf galaxy that has recently been absorbed by the Milky Way.

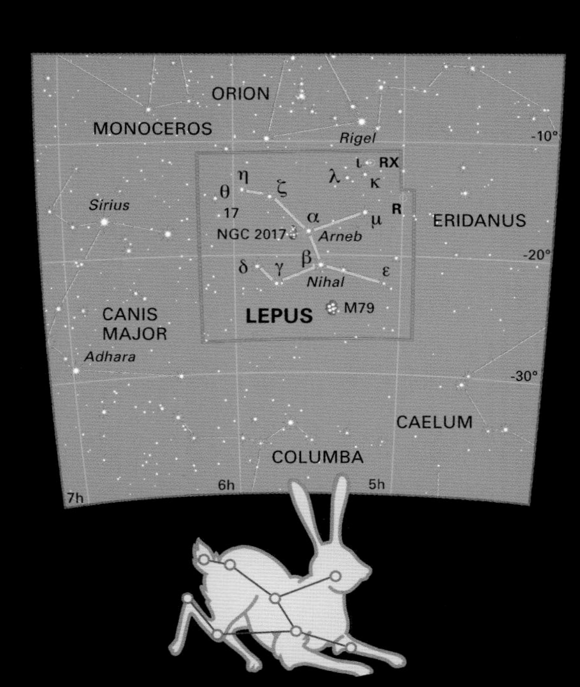

Safe Haven
Lepus. the celestial hare, crouches under the feet of Orion, like an animal trying to hide from its hunter. Orion's dogs, Canis Major and Canis Minor, lie nearby.

Columba The Dove

Fully visible 46°N–90°S

The faint constellation of Columba was invented by Dutch astronomer Petrus Plancius in 1592. Since Plancius was a biblical scholar, he probably intended the dove to represent the bird that Noah sent out from the Ark in search of dry land. However, others have linked it to a different dove, from classical myth, which Jason sent ahead of his ship the Argo to find a safe passage into the Black Sea. This may have been partly in Plancius's mind when he placed his dove so close to the constellation of Puppis, part of Argo.

FEATURES OF INTEREST

Alpha (α) Columbae (Phact) This star's name derives from the Arabic word for a collared dove. It is 170 light years from Earth and shines blue-white at magnitude 2.6.

Beta (β) Columbae (Wazn) The constellation's second brightest star is a yellow giant, 130 light years away, and shining at magnitude 3.1. Its Arabic name, Wazn, means "weight".

NGC 1851 This globular cluster is Columba's most prominent deep-sky object. It lies about 39,000 light years away and, at magnitude 7.1, is visible as a faint patch through binoculars or a small telescope.

Pyxis The Compass

Fully visible 52°N–90°S

Representing a magnetic compass (as opposed to Circinus, the draftsman's compasses), Pyxis is one of Nicolas Louis de Lacaille's technological constellations, added to the sky in the 1750s. In ancient times, its stars were probably incorporated into the great ship, Argo Navis.

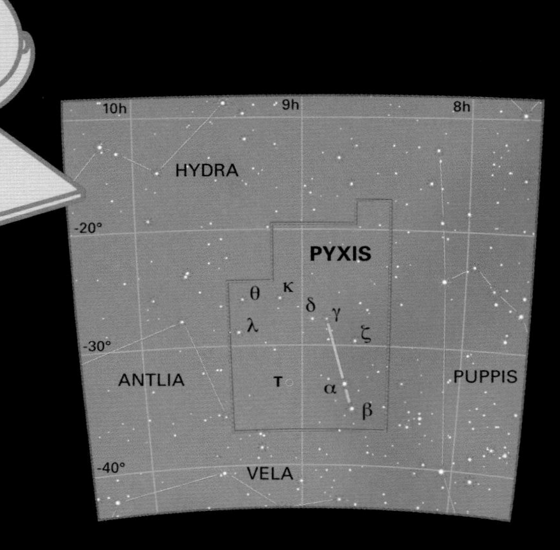

FEATURES OF INTEREST

Alpha (α) Pyxidis The brightest star, in the middle of the row of three linked stars, is a blue-white supergiant. It is 18,000 times more luminous than the Sun, yet it shines at a magnitude of only 3.7 because it is more than 1,000 light years distant.

Beta (β) Pyxidis In contrast to alpha, beta is a magnitude-4.0 yellow giant, 320 light years away.

T Pyxidis Most of the time, this variable star lies beyond the range of small telescopes, but occasionally it brightens dramatically to well within binocular range, and just below naked-eye visibility. The system is an unpredictable recurrent nova – a double star system in which a white dwarf is pulling material onto its surface from a larger neigbouring star. When the gases in the dwarf's atmosphere become sufficiently hot and dense, the star erupts in an enormous explosion.

Puppis The Stern

Fully visible 39°N–90°S

Puppis is one of three constellations that were once parts of the largest constellation in the sky, the huge ship Argo Navis. Puppis, representing the ship's stern, is the largest part. The stars of each section retained their original Greek letters, and in the case of Puppis the lettering now starts at Zeta (ζ) Puppis, a star that is also known as Naos.

FEATURES OF INTEREST

Zeta (ζ) Puppis (Naos) The splitting of Argo has left this blue giant as the brightest star in Puppis. It is also one of the hottest stars known, with a surface temperature six times that of the Sun. At some 14,000 light years away, it is only its vast distance that reduces Naos's brightness to magnitude 2.2.

L Puppis This line-of-sight double star consists of a blue-white star (L^1) 150 light years from Earth, and a red giant (L^2), 40 light years beyond it. L^1 shines at a steady magnitude 4.9, but L^2 is a pulsating variable, ranging in brightness between magnitude 2.6 and 6.2 over a 140-day cycle. Binoculars allow an easy comparison between the stars.

M47 This naked-eye object is a rich sight in binoculars. Slightly brighter than the neighbouring M46, it is 1,600 light years distant and the most impressive of several open star clusters in Puppis.

Vela The Sails

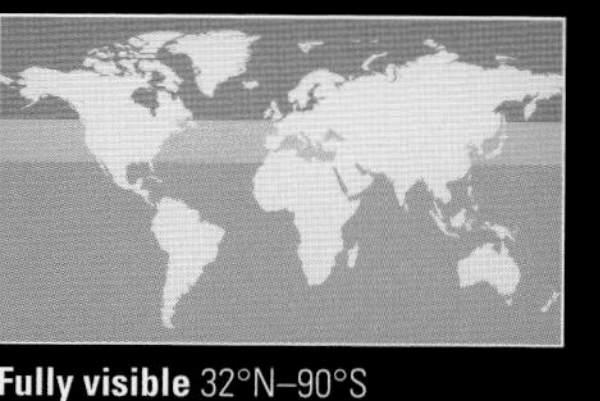

Fully visible 32°N–90°S

This large constellation represents the sail of the ship Argo Navis, and once formed a single huge constellation with Puppis and Carina. It outlines a dense region of the Milky Way that contains many interesting objects.

FEATURES OF INTEREST

Gamma (γ) Velorum (Suhail) Vela's brightest star is a multiple that contains the brightest known Wolf–Rayet star – a type of massive, superhot star that is blowing away its own outer layers with fierce stellar winds, thereby exposing its ultra-hot interior. The system shines at an overall magnitude of 1.8.

IC 2391 This beautiful star cluster is also called the Southern Pleiades. Just 400 light years away, its 30 or so naked-eye stars are a spectacular sight in binoculars.

Vela Supernova Remnant At the end of a massive star's life, it explodes as a supernova, shredding and dispersing its outer layers. The Vela Supernova Remnant (SNR), between Gamma and Lambda (λ) Velorum, results from one such explosion, 11,000 years ago. Its gas strands are diffuse and faint and are best seen through large telescopes or in long-exposure photographs.

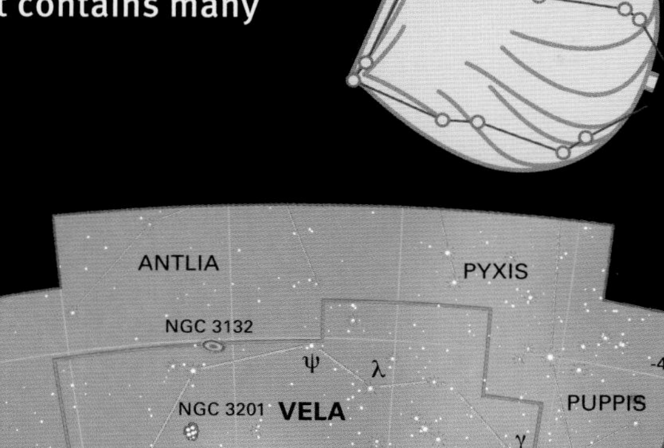

Under sail
Vela represents the mainsail of the *Argo*, the ship of Jason and the Argonauts, sailing through the southern sky in the quest for the golden fleece.

Even keel
Carina represents the keel and hull of the Argonaut's ship, the *Argo*. The blade of the steering oar is marked by Canopus, Carina's brightest star.

Carina The Keel

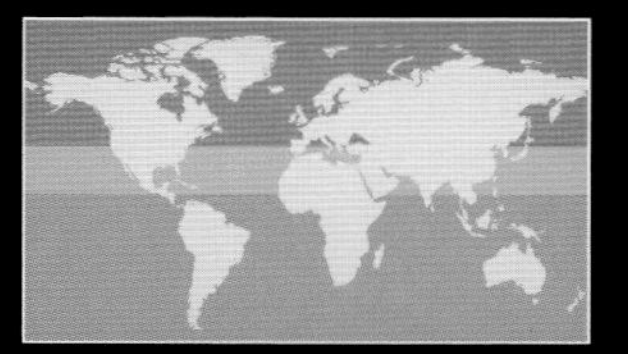

Fully visible 14°N–90°S

Marking the keel of the great ship Argo Navis, Carina lies right across the southern Milky Way, and contains many bright stars and interesting objects. As the most southerly of Argo's segments, the constellation is circumpolar (never setting) for much of the southern hemisphere.

FEATURES OF INTEREST

Theta (θ) Carinae (Canopus) The second brightest star in the entire sky, Canopus shines at a brilliant magnitude -0.7. Unlike Sirius, which outshines it purely because of its proximity to Earth, Canopus is a truly luminous star – a yellow-white supergiant some 310 light years away.

Carina Nebula (NGC 3372) This expansive emission nebula is a star-forming region 8,000 light years away. Four times the apparent size of the full Moon, it is visible to the naked eye as a bright patch in the Milky Way. The densest and brightest part of the nebula is around Eta (η) Carinae.

Eta (η) Carinae Normally shining at magnitude 6.7 in the heart of NGC 3372, this red supergiant is rapidly nearing the end of its life, and prone to violent outbursts that brighten it to naked-eye visibility. One such flare-up in the 19th century temporarily made it the second-brightest star in the sky. The star could destroy itself in a supernova at any time.

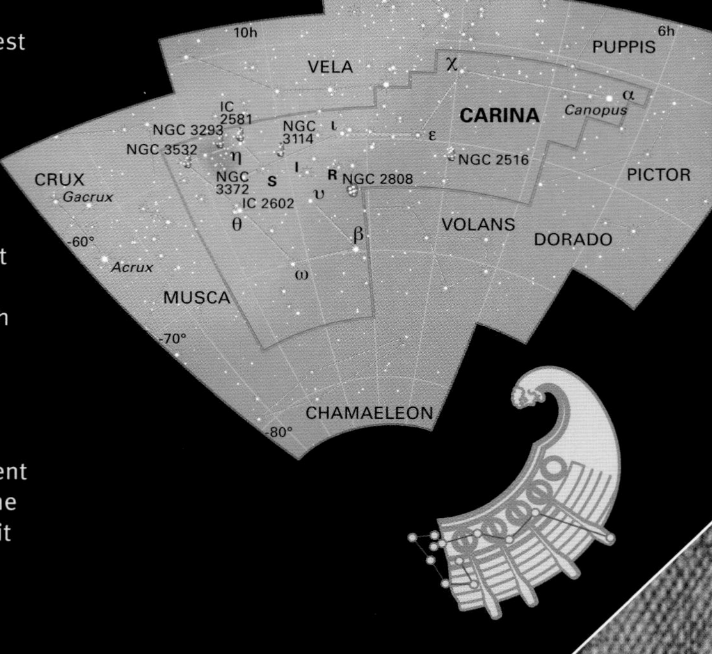

Crux The Southern Cross

Fully visible 25°N–90°S

The sky's smallest constellation is also one of its most distinctive thanks to its four bright stars. It was first mapped as a separate pattern by explorers in the early 1500s. Crux offers a useful pointer to the south celestial pole – following the line between alpha and gamma for just under five times its length reveals its location.

FEATURES OF INTEREST

Alpha (α) Crucis (Acrux) Marking the southern end of the cross, magnitude-0.8 Acrux is a blue-white double star, divisible in a telescope into two stars of roughly equal magnitude.

Beta (β) Crucis (Becrux) This rapidly varying blue-white star changes its brightness by less than 0.05 magnitude, either side of an average of 1.2, every 6 hours. It lies 350 light years from Earth.

NGC 475 5 This glorious star cluster, also known as the Jewel Box, appears to the naked eye as a single fuzzy star of magnitude 4.0. In fact, it is a cluster of stars 7,600 light years away. Binoculars reveal that it contains several dozen blue-white stars, with a contrasting red supergiant near the centre.

The Coal Sack This dark nebula, just 400 light years from Earth, is very noticeable because of the way it blocks out light from the dense Milky Way star fields behind it.

Musca The Fly

Fully visible 14°N–90°S

This is the most distinctive of several constellations invented by Dutch navigators Pieter Dirkszoon Keyser and Frederick de Houtman in the late 16th century. Its stars are relatively bright, and it is easily found by following the long axis of Crux, the Southern Cross, towards the south celestial pole.

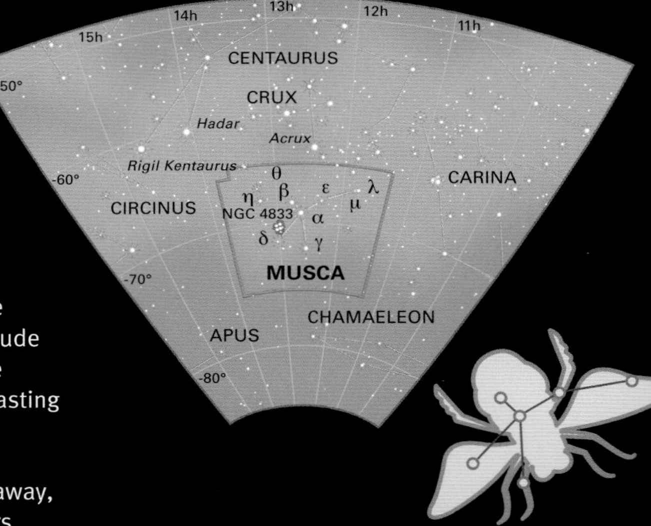

FEATURES OF INTEREST

Alpha (α) Muscae This magnitude-2.7 star is a blue-white giant, 305 light years away.

Beta (β) Muscae To the naked eye, beta is an almost identical twin of alpha. However, a small telescope will show that it is actually a double, containing two blue stars, of magnitudes 3.0 and 3.7, that orbit each other in 383 years. The system lies some 310 light years from Earth.

Theta (θ) Muscae This double star consists of a blue supergiant of magnitude 5.7 orbited by a star of magnitude 7.3. The faint companion, visible in binoculars, is a rare Wolf–Rayet star, a fierce white star – so hot that it is blasting its own outer layers away into space and ageing at an accelerated rate.

NGC 4833 This globular cluster, 18,000 light years away, shines at magnitude 6.5 and is easily seen in binoculars.

Circinus The Compasses

Fully visible 19°N–90°S

Supposedly resembling a pair of surveyor's compasses (as opposed to Pyxis, the magnetic compass), this faint triangle of stars is another of French astonomer Nicolas de Lacaille's 18th-century additions to the sky. It is easy to locate, however, as it lies beside Centaurus's brightest stars.

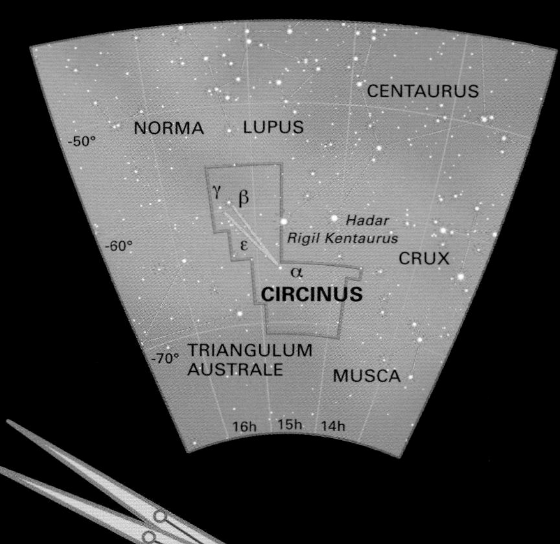

FEATURES OF INTEREST

Alpha (α) Circini This white star of magnitude 3.2 has a faint companion of magnitude 8.6 that can be seen through a small telescope. The system is 65 light years away.

Gamma (γ) Circini This double can be separated only with a medium-sized telescope. It consists of blue and yellow stars with magnitudes 5.1 and 5.5 respectively, orbiting each other 500 light years from Earth.

Theta (θ) Circini This is a variable, double star: its parts are too close to separate visually, but we know the system is young, since one of the stars still fluctuates unpredictably, causing theta's brightness to vary between magnitudes 5.0 and 5.4.

The Circinus Galaxy Despite lying at a distance of only 13 million light years, this active galaxy was recently discovered, below Hadar, concealed by the Milky Way.

Norma The Level

Fully visible 29°N–90°S

Sandwiched between the brighter stars of Scorpius and Lupus lies this faint triangle of stars. French astronomer Nicolas Louis de Lacaille decided in the 1750s that it resembled a surveyor's level or set-square. He originally called it Norma et Regula, or the "square and ruler".

FEATURES OF INTEREST

Gamma (γ) Normae Of the two components of this double star, γ^2 is a giant of magnitude 4.0, some 125 light years away, while γ^1 is a much more distant supergiant, 1,500 light years away but still shining at magnitude 5.0. Both are yellow, and they show how much the true brightness of apparently similar stars can vary.

Iota (ι) Muscae This star is 220 light years from Earth and appears double through a small telescope, with a magnitude-4.6 primary orbited by a fainter star of magnitude 8.1. Much larger telescopes can split the primary again into twin stars that orbit one another in 27 years. This makes the system a triple star.

NGC 6087 This open cluster of about 40 stars lies 3,000 light years away, and can be spotted with the naked eye. Its stars are mostly hot, young, and blue-white, but at its heart lies a yellow supergiant, S Normae.

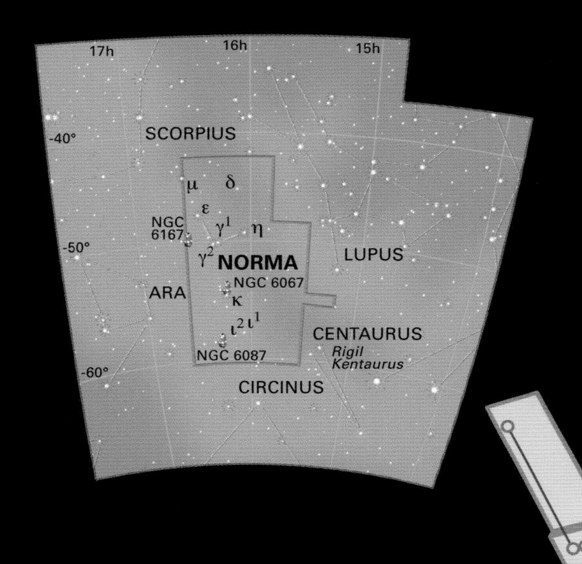

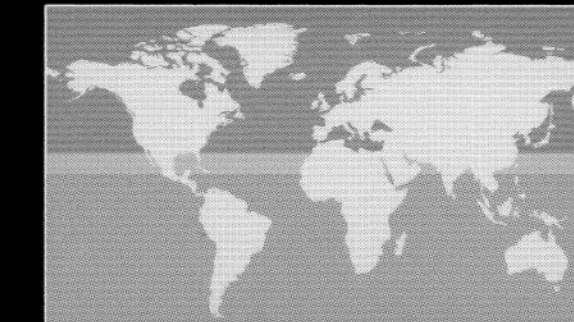

Right angle
Norma is a rather unremarkable southern constellation. Its most distinctive feature is a right-angled trio of three faint stars, which is somewhat difficult to identify among the rich Milky Way star fields.

Triangulum Australe The Southern Triangle

Fully visible 19°N–90°S

This distinctive pattern is easy to spot, but there are several claims to its invention. It was first recorded in *Uranometria*, the great star atlas compiled by Johann Bayer in 1603, but it may have been invented by the Dutch sailors Dirkszoon Keyser and de Houtman in the 1590s, or by a Dutch astronomer, Petrus Theodorus Embdanus, some decades earlier. Arab astronomers may also have named it independently.

FEATURES OF INTEREST

Alpha (α) Trianguli Australis The brightest star marks the triangle's southeast corner and is an orange giant of magnitude 1.9, lying 100 light years away.

Beta (β) Trianguli Australis This white star is about 42 light years from Earth and shines at magnitude 2.9.

Gamma (γ) Trianguli Australis Although it has the same magnitude (2.9) as beta, gamma lies over 70 light years away, so it is significantly more luminous. In keeping with its higher luminosity, its surface is hotter, and blue-white.

NGC 6025 Though visible to the naked eye at magnitude 5.4, this open cluster is best seen through binoculars.

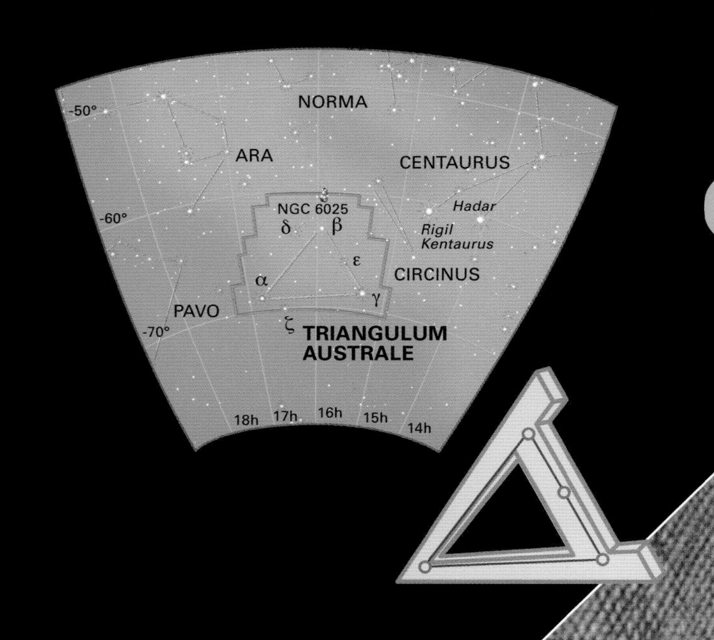

Southern triplet
Triangulum Australe is an easily recognized triangle of stars, lying in the Milky Way near brilliant alpha (α) and beta (β) Centauri.

Ara The Altar

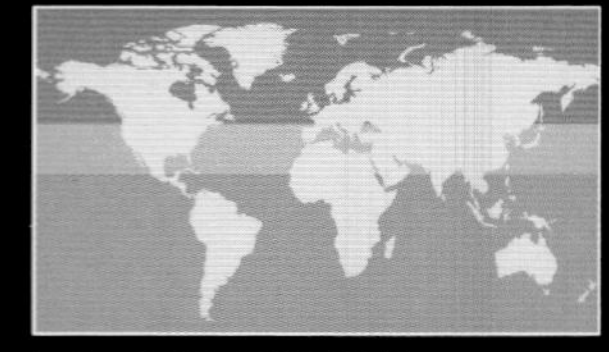

Fully visible 22°N–90°S

Although it lies in the far south of the sky, Ara originated with the ancient Greeks, who saw it as the altar on which the gods swore their oaths. Its pattern is obscure, but nevertheless is easy to locate to the south of Scorpius, and it is crossed by a rich band of the Milky Way's star fields.

FEATURES OF INTEREST

Alpha (α) Arae This blue-white star of magnitude 3.0 lies some 460 light years from Earth.

Gamma (γ) Arae This is one of the most luminous stars in our region of the galaxy, shining with the brilliance of 32,000 Suns. However, at a distance of 1,100 light years, it only reaches magnitude 3.3 in Earth's skies.

NGC 6193 This is a bright open cluster of stars, easily spotted with the naked eye, being over half the apparent width of the full Moon. It features a central blue-white giant that just reaches naked-eye visibility (magnitude 5.7) in its own right. The cluster is 4,000 light years from Earth, and still embedded in remnants of the gas from which it was born.

NGC 6397 A relatively nearby globular cluster, at a distance of just 7,200 light years, NGC 6397 is easily seen with binoculars.

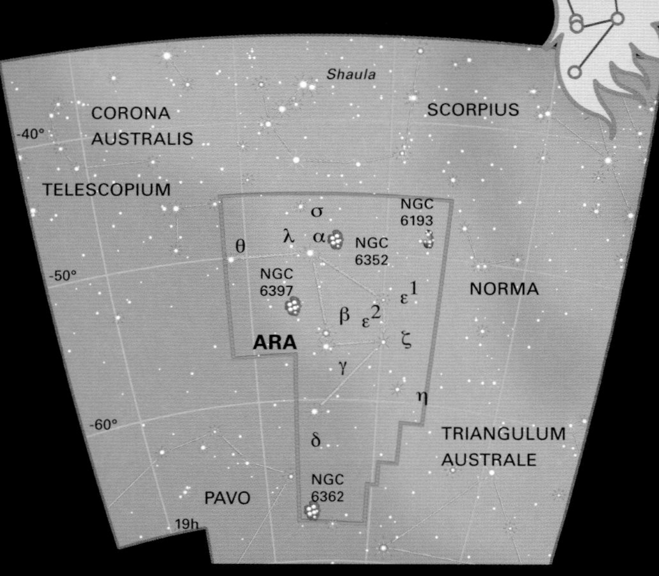

Incense burner
Ara, the celestial altar, is oriented with its top facing south. Incense burning on the altar might give off the "smoke" of the Milky Way above it.

Corona Australis The Southern Crown

Fully visible 44°N–90°S

Although the pattern of this celestial coronet is not as well-spaced as its northern equivalent, Corona Borealis, the southern crown is easy to recognize, just below the central "teapot" pattern of stars in Sagittarius. The constellation borders on the rich star clouds of the Milky Way.

FEATURES OF INTEREST

Alpha (α) Coronae Australis This magnitude-4.1 star is white in colour and 140 light years from Earth.

Beta (β) Coronae Australis This yellow giant shines at magnitude 4.1, with the same brightness in Earth's skies as alpha. However, in reality it is considerably further away, at a distance of 510 light years. This means it is actually 13 times more luminous than alpha. With a radius half the size of Mercury's orbit, it is 730 times more luminous than the Sun.

Gamma (γ) Coronae Australis Both the stars in this binary system are bright enough to be seen with the naked eye (magnitudes 4.8 and 5.1). However, a small telescope is still needed to separate them.

NGC 6541 Hovering just below naked-eye visibility, this globular cluster is 22,000 light years from Earth.

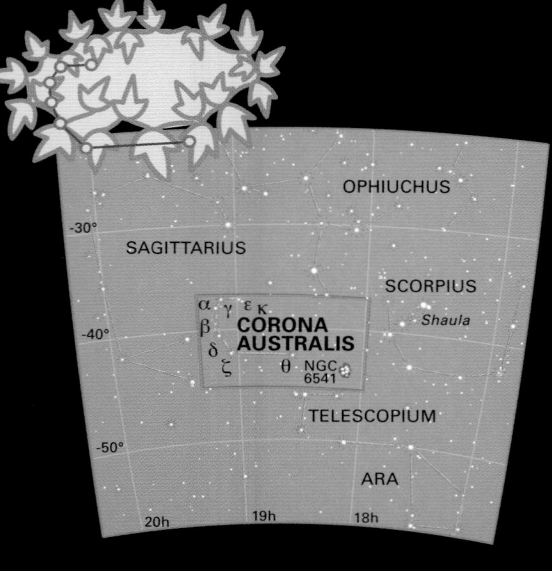

Southern arc
Corona Australis lies under the feet of Sagittarius. It is an attractive arc of stars that represents a crown or laurel wreath.

The Telescope

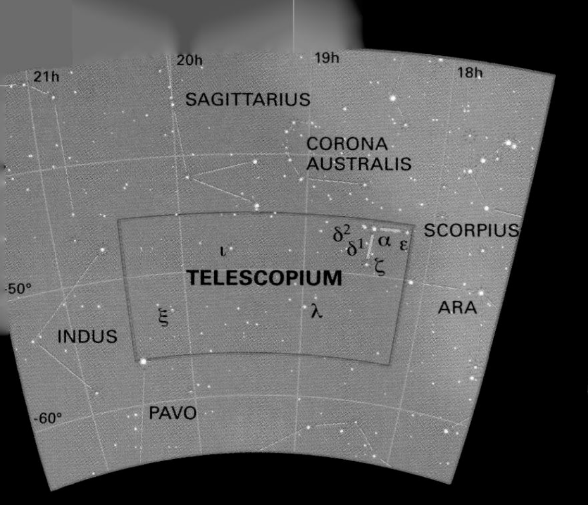

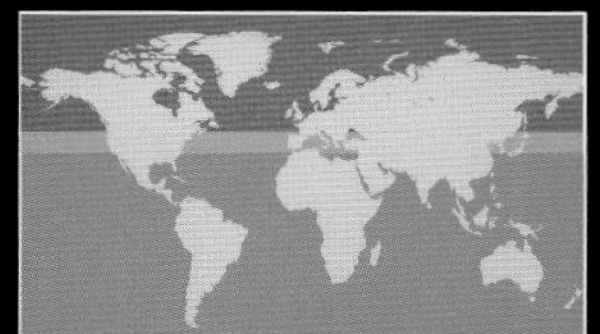

Fully visible 33°N–90°S

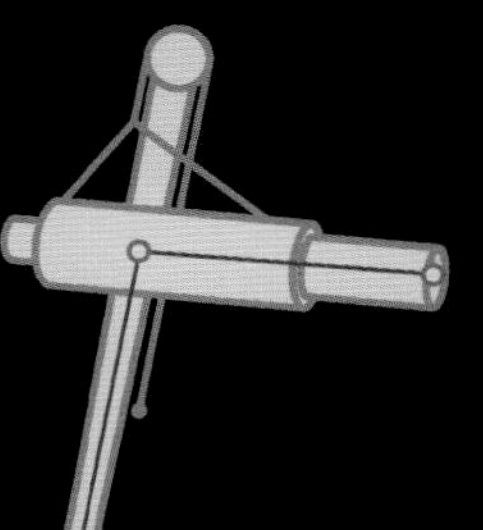

One of the least recognizable constellations, Telescopium seems to have been made simply by drawing a line around an arbitrary area of sky. An invention of French astronomer Nicolas Louis de Lacaille during his South African observing tour of the 1750s, the constellation does at least have an interesting history – when Lacaille compiled the pattern, he "stole" stars from several nearby constellations, including Sagittarius, Scorpius, Ophiuchus, and Corona Australis. When the constellations were standardized in 1929, these stars were returned to their rightful owners, and Telescopium was left in its current state.

FEATURES OF INTEREST

Alpha (α) Telescopii This blue-white star is around 450 light years from Earth and shines at magnitude 3.5.

Delta (δ) Telescopii Binoculars will reveal that this star is a line-of-sight double, consisting of two unrelated blue-white stars. These stars are 650 and 1,300 light years away. They are of roughly equal brightness at around magnitude 5.0. They can even be divided with good eyesight.

Indus The Indian

Fully visible 15°N–90°S

This constellation, representing a Native American, is an invention of Dutch navigators Frederick de Houtman and Pieter Dirkszoon Keyser, who made long voyages in the southern hemisphere during the 1590s. They made the first record of the far southern stars at the request of Dutch astronomer Petrus Plancius, who had already added several new constellations to northern skies.

FEATURES OF INTEREST

Alpha (α) Indi This orange giant star, 125 light years from Earth, shines at magnitude 3.1.

Beta (β) Indi A slightly less luminous orange giant, this star is 15 light years closer than alpha, but still dimmer in Earth's skies, at magnitude 3.7.

Epsilon (ε) Indi This yellow star of magnitude 4.7 is a close stellar neighbour of the Sun, 11.2 light years away. It is orbited by a "brown dwarf" 45 times the mass of Jupiter. Though beyond the range of amateur telescopes, the brown dwarf is is a prime target for future telescopes that will search for Earth-like planets around other stars.

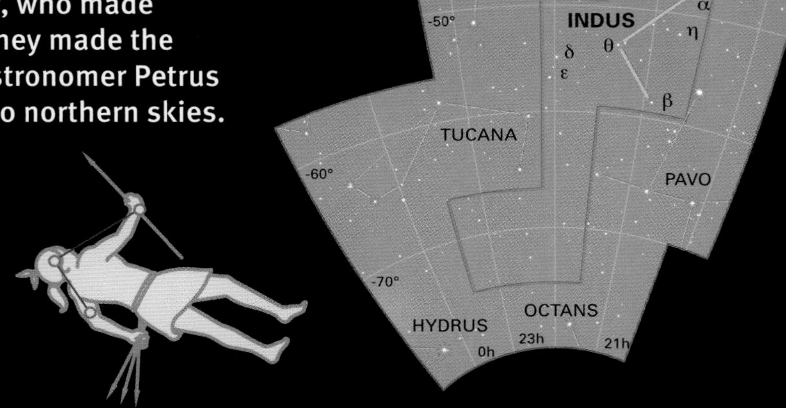

Grus The Crane

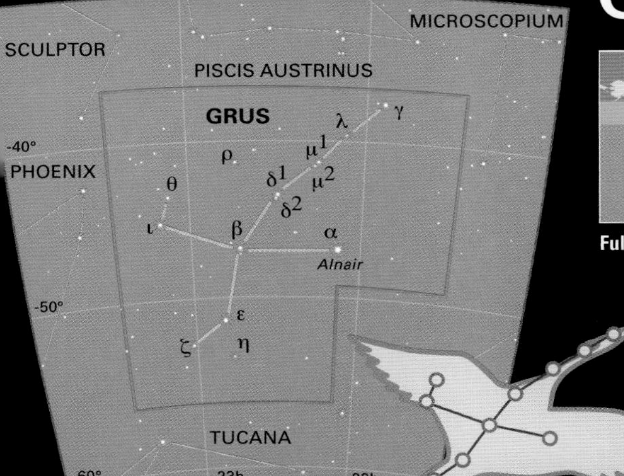

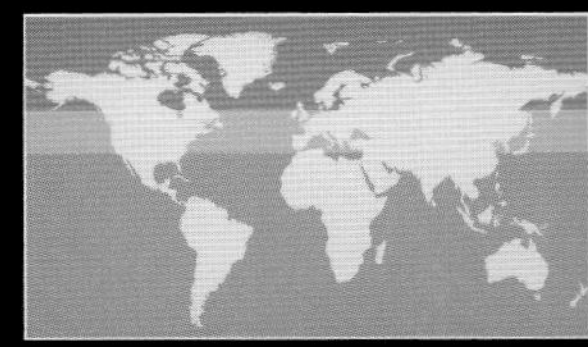

Fully visible 33°N–90°S

Grus is one of several bird-shaped constellations added to the southern sky in the 1590s by the Dutch explorers Pieter Dirkszoon Keyser and Frederick de Houtman and was later immortalized by Johann Bayer in his Uranometria star atlas of 1603. The stars of the crane's neck form a chain between the Small Magellanic Cloud in Tucana and Fomalhaut in Piscis Austrinus.

FEATURES OF INTEREST

Alpha (α) Gruis (Alnair) This blue-white star, whose name means "the bright one" in Arabic, is 65 light years away and shines at magnitude 1.7.

Beta (β) Gruis Beta is a red giant, 170 light years from Earth. It is variable, oscillating unpredictably between magnitudes 2.0 and 2.3 as it swells and shrinks.

Delta (δ) Gruis Naked-eye observers can usually tell that this star is a double, but this is in fact just a line-of-sight effect. The two components are yellow and red giants of magnitudes 4.0 and 4.1, 150 and 420 light years away respectively. Together with Mu (μ) Gruis, this star appears along the extended neck of Grus.

Phoenix falling
The stars of Phoenix sink toward the western horizon in the morning sky, with Grus below it. North is to the right in the photograph.

Phoenix The Phoenix

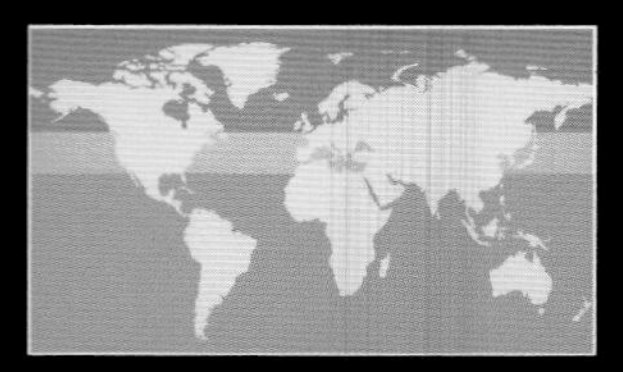

Fully visible 32°N–90°S

Representing the mythical firebird that regenerates from its own ashes, Phoenix is an indistinct group, although easily located because it lies beside the brilliant Achernar. Arabian astronomers named the pattern after a boat moored on the river Eridanus.

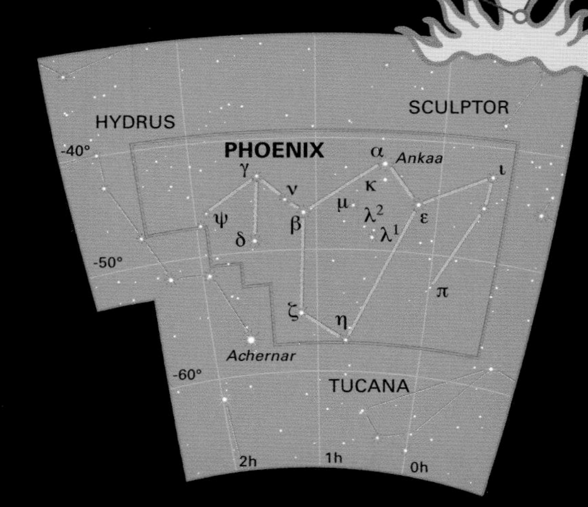

FEATURES OF INTEREST

Alpha (α) Phoenicis (Ankaa) Located 88 light years from Earth,this yellow giant shines at magnitude 2.4.

Beta (β) Phoenicis Although to the naked eye this appears to be a yellow star of magnitude 3.3, a medium-sized telescope will show that it is actually double, consisting of twin yellow stars of magnitude 4.0, 130 light years away.

Zeta (ζ) Phoenicis This interesting quadruple star lies 280 light years away. Its brightest component shines for most of the time at magnitude 3.9, but dips briefly to 4.4 every 40 hours. This is because it is an eclipsing binary – a close pair of stars that pass in front of each other once in every orbit. A small telescope will reveal a third star, of magnitude 6.9, while a larger instrument will show the fourth member of the system, fainter and much closer to the primary.

Tucana The Toucan

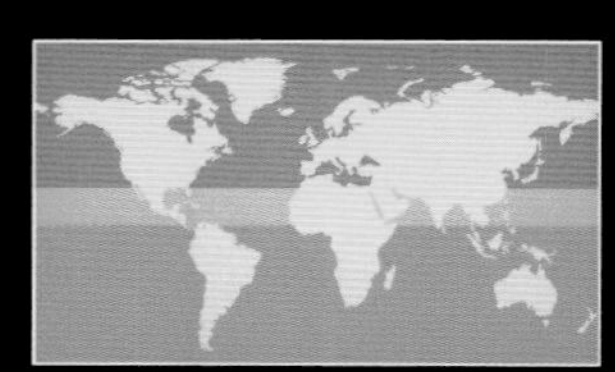

Fully visible 14°N–90°S

Invented, like most of the bird constellations, by Dutch navigators Pieter Dirkszoon Keyser and Frederick de Houtman, Tucana is an indistinct group of stars to the west of brilliant Achernar in Eridanus. However, it contains two outstanding objects of interest to any amateur astronomer.

FEATURES OF INTEREST

Small Magellanic Cloud The SMC is the smaller of the Milky Way's two major satellite galaxies, first recorded by Portuguese explorer Ferdinand Magellan in around 1520. It lies 210,000 light years from Earth and orbits our galaxy every 1.5 billion years. The SMC is easily visible to the naked eye, looking like a detached region of the Milky Way itself, but binoculars will reveal rich fields of stars, dust, and star-forming nebulae.

47 Tucanae Next to the SMC in the sky, but actually a foreground object within the Milky Way, this beautiful globular cluster is classified as a star but in fact contains hundreds of thousands of stars in a ball about 120 light years across. It is 13,500 light years away but is easily visible as a fuzzy star with the naked eye. Small telescopes can pick up single stars around this cluster's edge.

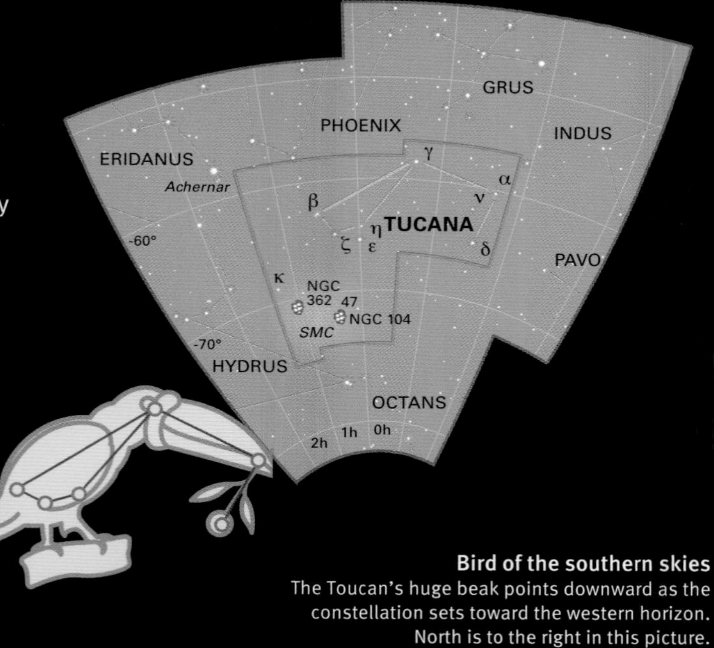

Bird of the southern skies
The Toucan's huge beak points downward as the constellation sets toward the western horizon. North is to the right in this picture.

Hydrus The Little Water Snake

Fully visible 8°N–90°S

While the similarly named Hydra is the longest constellation, Hydrus is compact and nestles farther to the south. Its stars are of middling magnitude and their pattern indistinct, but the constellation is easy to locate because alpha – the snake's "head" – lies near to the brilliant Achernar.

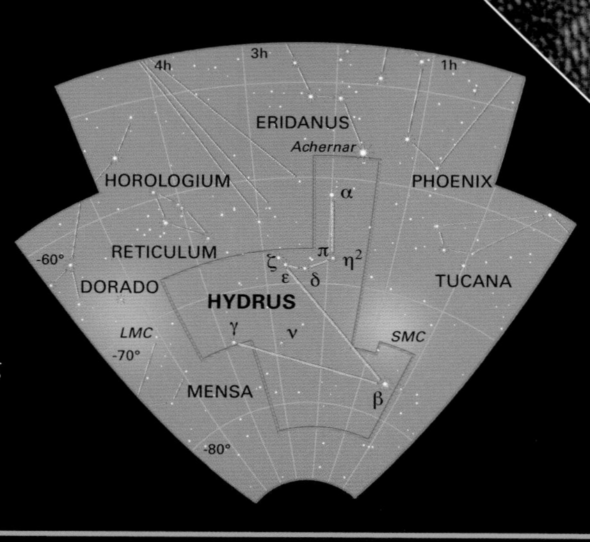

FEATURES OF INTEREST

Alpha (α) Hydri At magnitude 2.9, this white star, 78 light years from Earth, is the constellation's brightest member.

Beta (β) Hydri This Sun-like yellow star in the snake's "tail" is just 21 light years away and shines at magnitude 2.8.

Pi (π) Hydri This wide double of unrelated red giants can be split readily with binoculars. Pi-1 (π^1) is 740 light years away, while Pi-2 (π^2) lies closer, at a distance of 470 light years.

VW Hydri **Lying near Gamma (γ)** Hydri, VW is an interesting star whose variations are easy to follow in a small or medium-sized telescope. The star is actually a recurrent nova system – a binary containing a white dwarf that pulls material off its companion, which occasionally ignites in a fire storm on its surface. VW erupts roughly once a month, brightening from magnitude 13 to 8 in just a few hours, then fading away over several days.

Horologium The Pendulum Clock

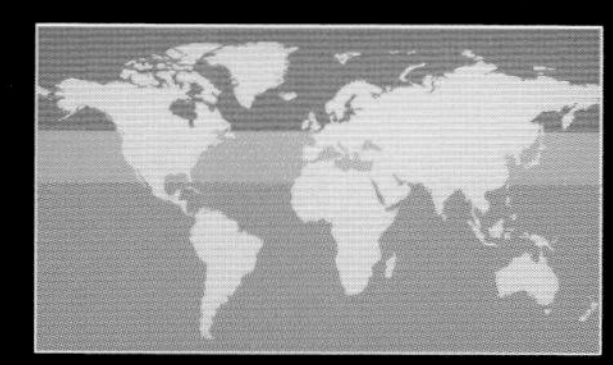

Fully visible 23°N–90°S

This faint constellation is one of Nicolas Louis de Lacaille's inventions, and, like many of them, it is a fairly arbitrary group of faint, scattered stars. Lacaille intended it to represent a clock, and it is often drawn with its pendulum suspended at alpha, and swinging back and forth between lambda and beta.

FEATURES OF INTEREST

Alpha (α) Horologii The constellation's brightest star is a magnitude-3.9 yellow giant, 180 light years from Earth.

NGC 1261 This globular cluster is one of the more distant of its type, 44,000 light years away from Earth. The combined light of this huge ball of stars reaches Earth at magnitude 8.0, making it a good target for binoculars.

NGC 1512 This barred spiral galaxy is about 30 million light years away and is nearly 70,000 light years across, as large as the Milky Way galaxy. Its bright centre has a magnitude of 10, and is visible through a small telescope, while detailed observations have shown that the centre is surrounded by a huge ring of infant star clusters. This region of star formation is about 2,400 light years across.

Reticulum The Net

Fully visible 23°N–90°S

This faint but distinct diamond-shape of stars, invented by the French astronomer Nicolas Louis de Lacaille, lies a little way to the south of the brilliant star Canopus. The literal meaning of its Latin name is the "net", but it is actually supposed to represent a reticle – the set of crosshairs in the eyepiece of a telescope and some other scientific instruments.

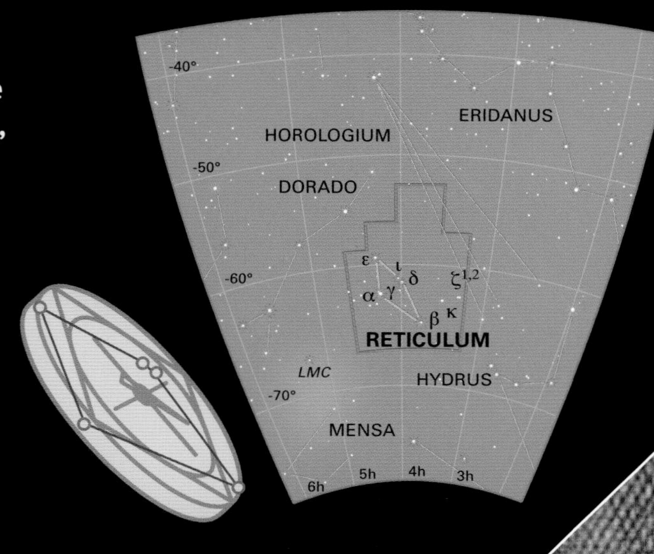

FEATURES OF INTEREST

Alpha (α) Reticuli The constellation's brightest star at magnitude 3.4, alpha is a yellow giant located 135 light years away from Earth.

Beta (β) Reticuli Beta is an orange giant some 78 light years away, and shines at magnitude 3.9.

Zeta (ζ) Reticuli This double star, easily resolved through binoculars, consists of twin yellow stars – Zeta-1 (ζ^1) and Zeta-2 (ζ^2). These have magnitudes 5.2 and 5.9 respectively and lie 39 light years away. The chemical composition of this system suggests that the stars in this very wide binary may be up to 8 billion years old – far more ancient than the Sun – and astronomers are eager to search for possible planets in orbit around them.

Dividing line
Pictor consists of little more than a crooked line of stars between brilliant Canopus (in Carina), seen here on the left, and the Large Magellanic Cloud.

Pictor The Painter's Easel

Fully visible 26°N–90°S

Located just to the west of Canopus, the second brightest star in the sky, Pictor is one of Lacaille's 18th-century constellations. Typically for one of his inventions, it is a group of faint stars with no obvious resemblance to the object – in this case, an artist's easel – that they supposedly represent. However, Pictor is redeemed by the presence of two very interesting stars.

FEATURES OF INTEREST

Beta (β) Pictoris Uninspiring at first glance, this white star of magnitude 3.9, situated 63 light years away, nevertheless has an interesting secret that it reveals in infrared (heat) radiation. The star is surrounded by a broad disc of planet-forming gas and dust, and recent studies have shown that something close to the star (most likely a newborn planet) is warping the disc out of shape. The planets of our Solar System are believed to have developed from a similar disc that existed shortly after its formation.

Delta (δ) Pictoris This is an eclipsing binary star – a pair of stars too close to separate with even the most powerful telescope. They give themselves away by periodic dips in their brightness (from magnitude 4.7 to 4.9) as one passes in front of the other every 40 hours. They are 2,400 light years from Earth.

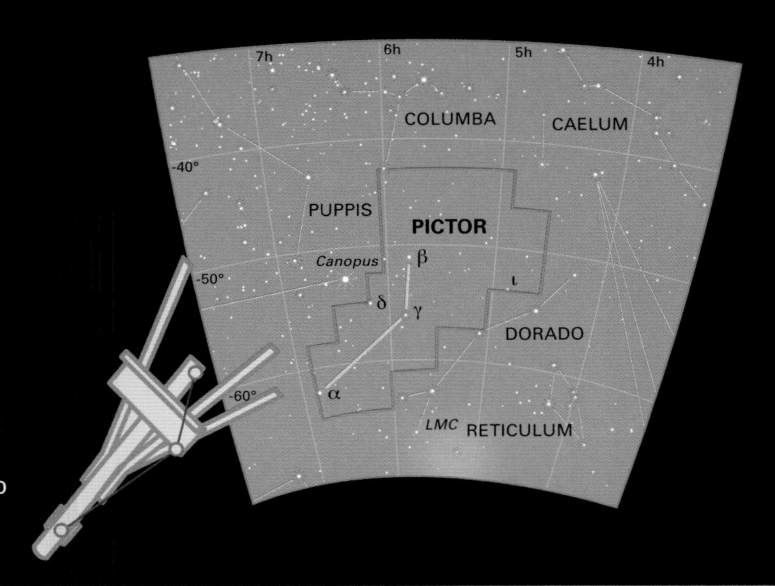

Dorado The Goldfish

Fully visible 20°N–90°S

Invented by Dutch navigators Dirkszoon Keyser and de Houtman in the 1590s, Dorado is also called the Swordfish. Its stars form a faint chain near to Canopus. The brightest is magnitude 3.3, but Dorado contains an object far more impressive than its stars.

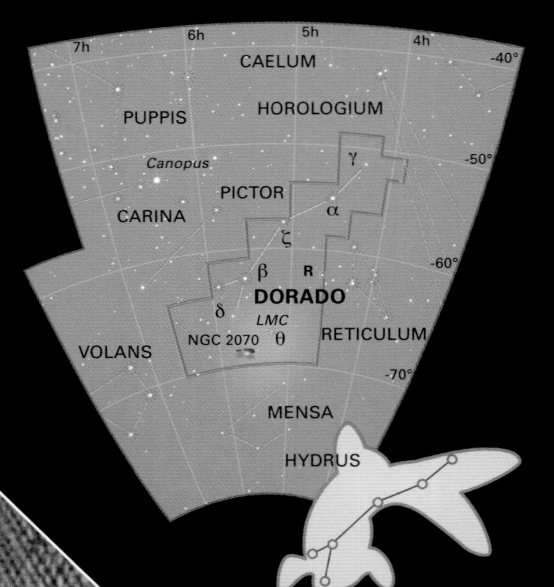

FEATURES OF INTEREST

Large Magellanic Cloud (LMC) The LMC is named after Portuguese explorer Ferdinand Magellan, who recorded it in the early 1520s, but it has been known to cultures of the southern hemisphere since prehistory. Arab astronomers in the 10th century named it Al Bakr, the "white ox". It is an irregular satellite galaxy of our own Milky Way, some 150,000 light years away, and is easily visible to the naked eye, although a small telescope gives the best views of its numerous star fields and nebulae.

Tarantula Nebula (NGC 2070) The spectacular Tarantula Nebula in the Large Magellanic Cloud is visible as a fuzzy star to the naked eye, and it also bears the stellar name 30 Doradus. But at 800 light years across, it is one of the largest star-forming regions known. A cluster of hot blue-white supergiant stars, known as R136, illuminates it from the inside.

Heading south
Dorado swims through the southern skies, apparently on its way to the south celestial pole. Although known as the goldfish, the constellation in fact represents the dolphinfish found in tropical waters, not common aquarium and pond fish.

Fish in flight
The fyling fish leaps into the evening sky above the eastern horizon. Beneath it are the Milky Way and the stars of Carina and Vela, with the False Cross at the left of the image.

Volans The Flying Fish

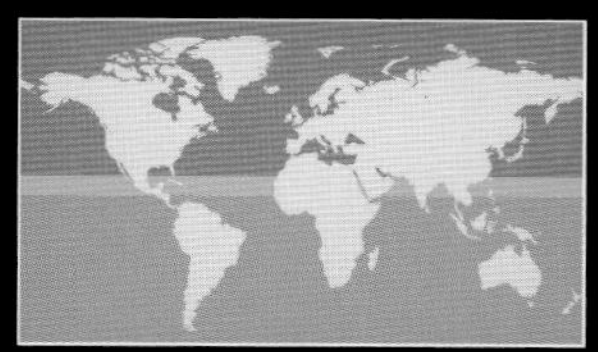

Fully visible 14°N–90°S

The 16th-century Dutch explorers Peter Dirkszoon Keyser and Frederick de Houtman named many of their discoveries after birds, but Volans is an exception, because it takes its inspiration from the bizarre flying fish of the Indian Ocean. The constellation is fairly faint and indistinct, but it is easily found because it lies between the bright stars of Carina and the south celestial pole.

FEATURES OF INTEREST

Gamma (γ) Volantis The constellation's brightest star has the wrong Bayer letter due to historical accident. It is a double that can easily be split through a small telescope to reveal a golden star of magnitude 3.8 and a yellow-white companion of magnitude 5.7. Both are 200 light years away from Earth.

Epsilon (ε) Volantis This is another interesting double star, although it is not as colourful as Gamma. The blue-white primary is 550 light years away and shines at a magnitude of 4.4. It has a magnitude-8.1 companion visible only through a small telescope.

NGC 2442 A larger telescope is needed to view this face-on barred spiral galaxy, 50 million light years away from Earth. Nevertheless, it is a beautiful sight, with spiral arms extending out to an elegant "S" shape from a pronounced central bar.

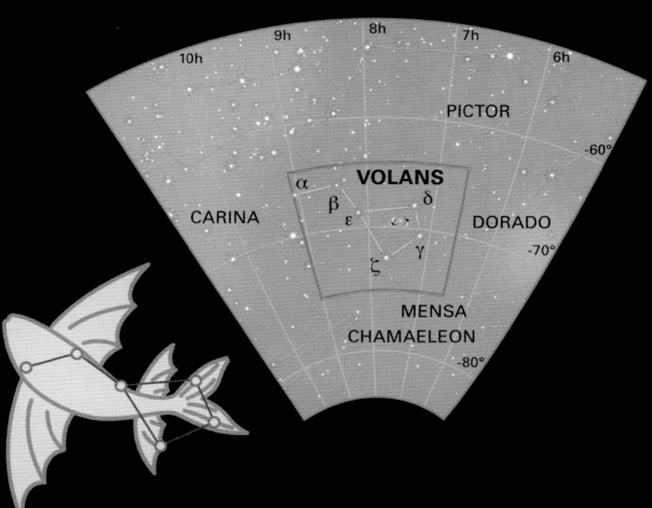

Mensa The Table Mountain

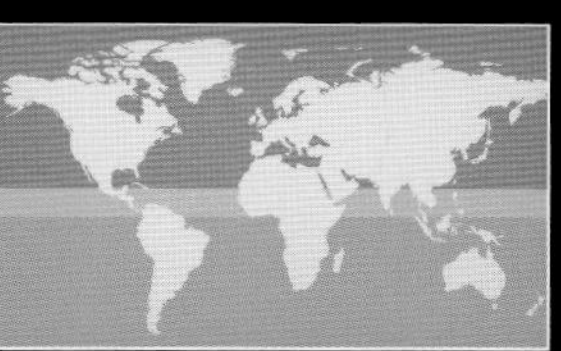

Fully visible 5°N–90°S

French astronomer Nicolas Louis de Lacaille, observing the southern skies from Cape Town in the early 1750s, named this constellation in honour of the distinctive Table Mountain that overlooks the city. Mensa is one of the faintest constellations in the sky, with no stars brighter than magnitude 5.0. However, it is fairly straightforward to locate, as it lies between the south celestial pole and the Large Magellanic Cloud (LMC) in Dorado. In fact, the southern reaches of the LMC cross over the border into Mensa itself.

FEATURES OF INTEREST

Alpha (α) Mensae The constellation's brightest star reaches a magnitude of only 5.1. It is an average yellow, Sun-like star, relatively close to Earth at a distance of 30 light years.

Beta (β) Mensae While beta is also yellow, and just slightly fainter than alpha at magnitude 5.3, it is a very different type of star. Astronomers have measured its distance at 300 light years, meaning that it must be a yellow supergiant, 100 times more luminous than alpha.

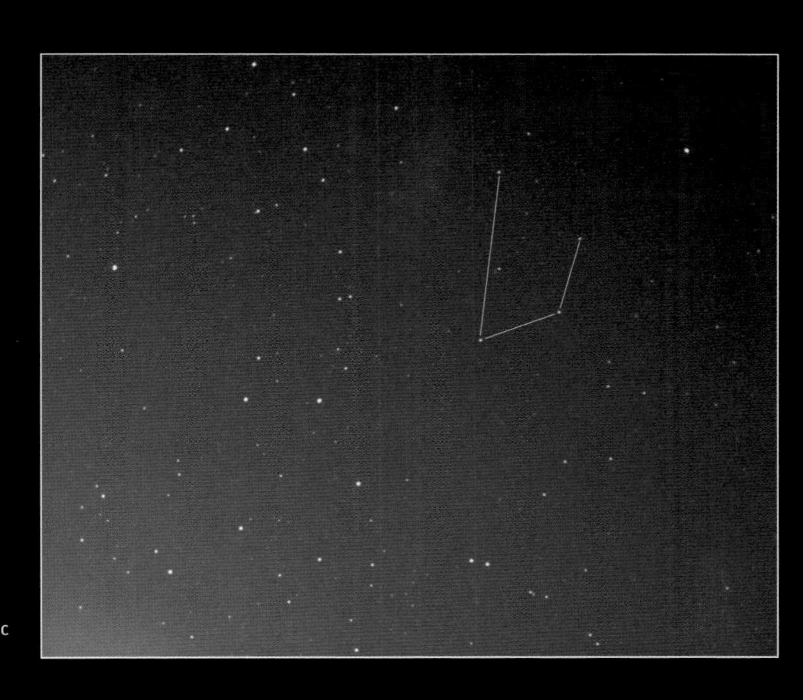

Table top
The far-southern constellation Mensa appears in this photograph above pink-tinged clouds in the dawn sky. Its main point of interest is a part of the Large Magellanic Cloud that overlaps into it from nearby Dorado.

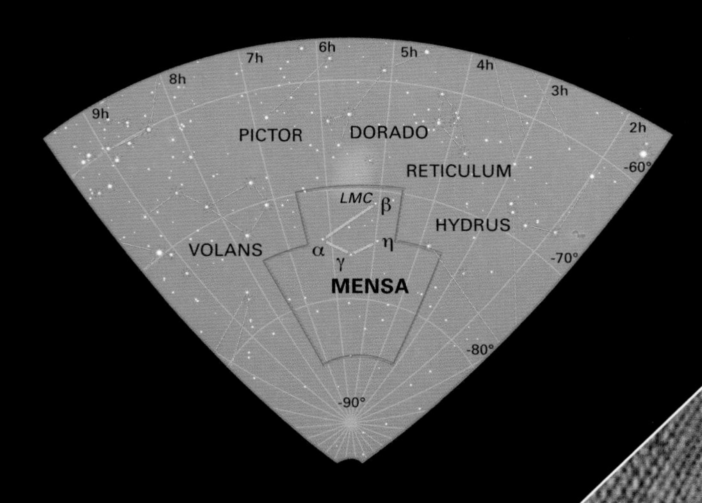

Camouflage artist
Chamaeleon lies close to the south celestial pole, which is to the left of it in this picture. To the north of this constellation are found the rich Milky Way star fields of Carina.

Chamaeleon The Chameleon

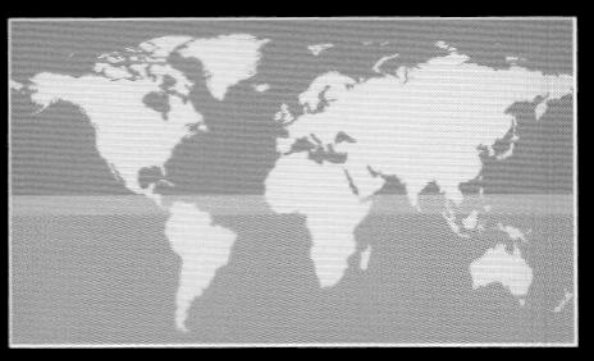
Fully visible 7°N–90°S

This skewed diamond of stars is supposed to represent a chameleon, but bears little resemblance to the lizard. It first appeared in Johann Bayer's Uranometria star atlas of 1603, and is probably another invention of the Dutch explorers Dirkszoon Keyser and de Houtman. It remains above the horizon at all times for every inhabited part of the southern hemisphere, but has few interesting objects.

FEATURES OF INTEREST

Alpha (α) Chamaeleontis This blue-white star shines at magnitude 4.1, and lies at a distance of 65 light years from Earth.

Delta (δ) Chamaeleontis Delta is a line-of-sight double star, easily separated in binoculars. Delta-1, the closer of the two stars, is an orange giant of magnitude 5.5, 360 light years away from Earth. Delta-2 is brighter but more distant, located 780 light years away and shining at magnitude 4.4.

NGC 3195 This ring-like planetary nebula, formed as a dying Sun-like star puffs its outer layers into space, shines at magnitude 10. Of similar apparent size to Jupiter, it is relatively faint and requires a medium-sized telescope to be seen. It is the most southern of all the planetary nebula in the sky, and remains completely invisible to all northern observers.

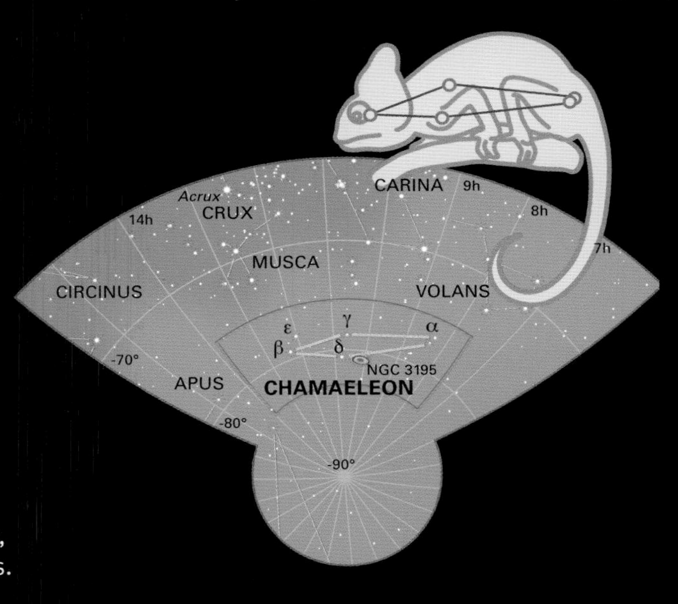

Apus The Bird of Paradise

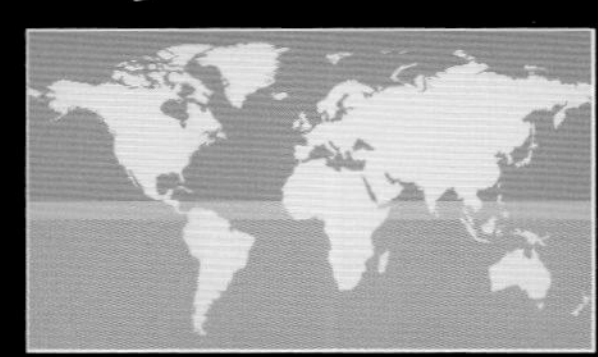
Fully visible 7°N–90°S

Pieter Dirkszoon Keyser and Frederick de Houtman named this constellation after the dazzling birds of paradise they saw during their explorations of New Guinea in the 1590s. Close to the south celestial pole, it is permanently visible in nearly all of the southern hemisphere. Despite its exotic name, the constellation is disappointingly obscure, with only a faint pattern of stars.

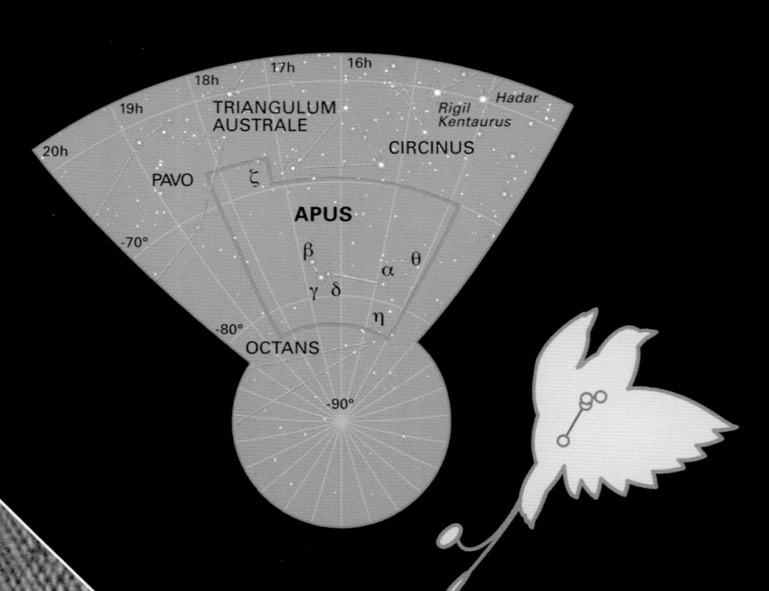

FEATURES OF INTEREST

Alpha (α) Apodis The constellation's luminary (brightest star) at a magnitude of 3.8, this orange giant is located about 230 light years from Earth.

Delta (δ) Apodis The most interesting star in Apus is this double, consisting of orange giants with magnitudes 4.7 and 5.3. The two stars orbit each other some 310 light years from Earth, and can easily be separated with binoculars – and sometimes by a sharp pair of eyes.

Theta (θ) Apodis Theta is a variable star, with fluctuations in brightness that are easily followed through binoculars. It ranges between magnitudes 6.4 and 8.0 in a 100-day cycle.

Exotic bird
Apus, which is south of the distinctive Triangulum Australe, represents a bird of paradise but is a disappointing tribute to such an exotic bird.

Pavo The Peacock

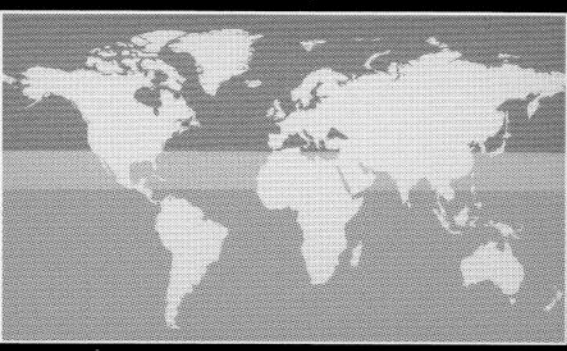

Fully visible 15°N–90°S

Another of the bird constellations added to the sky in the 1590s by Dutch navigators Frederick de Houtman and Pieter Dirkszoon Keyser, Pavo lies in a fairly featureless area of the sky, but is easily spotted because of its brightest star – Alpha Pavonis, the Peacock itself.

FEATURES OF INTEREST

Alpha (α) Pavonis (The Peacock) Alpha Pavonis is a blue-white giant star in the northeast corner of Pavo. It has a magnitude of 1.9 and is brilliant enough that, even at a distance of 360 light years from Earth, it is unmistakable.

Kappa (κ) Pavonis A yellow supergiant, 550 light years away, this is one of the brightest Cepheid variables in the sky. It expands and contracts in a 9.1-day cycle, varying in brightness between magnitudes 3.9 and 4.8 as it does so. These fluctuations can be followed with the naked eye.

NGC 6752 This globular cluster of stars, 14,000 light years away, is visible to the naked eye at magnitude 5.

NGC 6744 This barred spiral galaxy has its "face" open towards Earth and is located 30 million light years away.

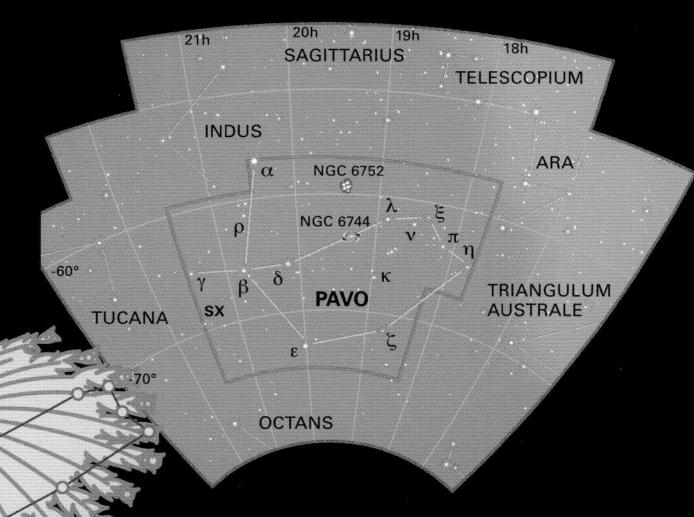

Celestial display
The constellation Pavo, the Peacock, is depicted fanning its tail across the southern skies, in imitation of a real-life peacock when attracting a male.

Octans The Octant

Fully visible 0°–90°S

While the north celestial pole is marked by the bright stars of Ursa Minor, the south polar constellation is faint and indistinct. It was invented by Nicolas Louis de Lacaille during his 18th-century observations from South Africa, and represents an octant – later supplanted by the sextant.

FEATURES OF INTEREST

Beta (β) Octantis A bright star in a dim constellation, beta is a white star that lies 110 light years from Earth. It shines at magnitude 4.1 and is outshone only by nu (n) at magnitude 3.8.

Gamma (γ) Octantis This is a chain of three stars, usually separable with the naked eye, whose members lie at different distances from Earth. Gamma-1 and Gamma-3 are both yellow giants, 270 and 240 light years away respectively, and shining at magnitudes 5.1 and 5.3. Between them lies Gamma-2, an orange giant of magnitude 5.7 at a distance of 310 light years.

Sigma (σ) Octantis The southern "pole star" is dim, white, and around 300 light years away. Its only noteworthy feature is its location within one degree of the south celestial pole, which makes it the nearest thing to a fixed reference point in the southern sky.

At the pole
Octans comprises only a scattering of faint stars. There is no bright star to mark the southern pole, which lies to the centre left in this picture.

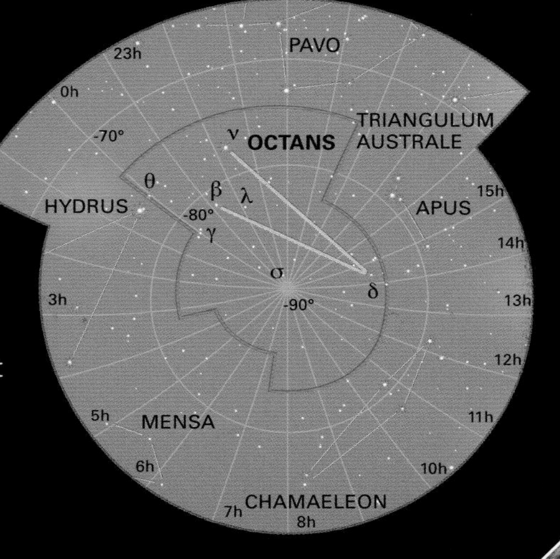

Monthly sky guide

Some constellations are always in our sky, while others appear and disappear during the course of the year. Your view of the night sky and the highlights that each month brings are described in this section. For each month, a Special Events table lists events that vary from year to year, while the positions of the naked-eye planets are shown on a locator chart.

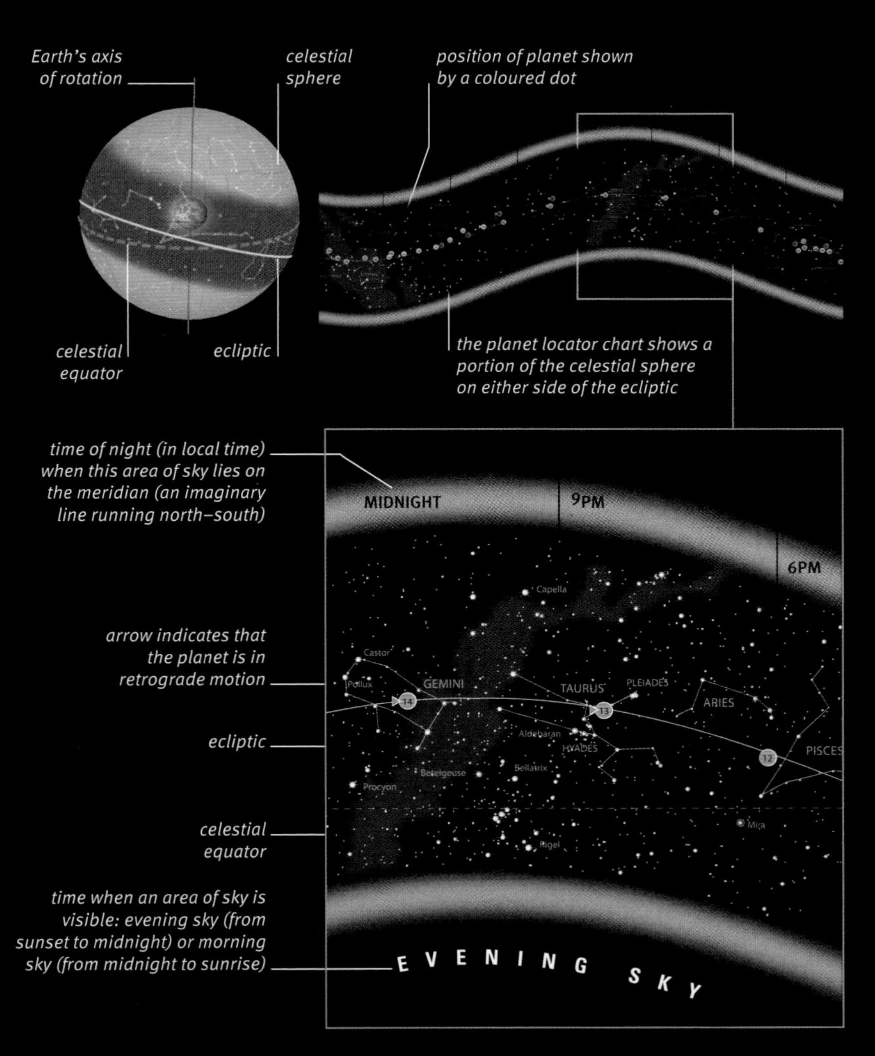

Planet locator charts
These charts show the positions of Mercury, Venus, Mars, Jupiter, and Saturn. They give the planets' positions in relation to the constellations along the ecliptic – the part of the sky in which they are always found. The view portrayed here and described in the accompanying text is for 10pm mid-month (11pm at the start of the month, and 9pm at the end). When daylight saving time is observed, add one hour to these times.

Using the charts
Each planet is represented by a differently coloured dot (the key is repeated in the guide for each month), and the number inside the dot refers to a particular year. Once you have identified the constellation in which the planet you are looking for can be found, use the planisphere to locate it in the night sky and the constellation charts (see pp.58–95) to get a more detailed view.

An ever-changing scene
There is always something new to see in the night sky. As Earth spins, the stars that are centre stage at the start of the evening are replaced by fresh ones from the east, which in turn sink to the west before the night is over.

January

The impressive constellation of Orion dominates the scene this month. The hunter's recognizable figure, with raised club and lion pelt, is easy to spot; and in his sword is the celebrated Orion Nebula. His two hunting dogs, Canis Major and Canis Minor, are close by. Canis Major includes Sirius, the brightest star of all in Earth's night sky.

SPECIAL EVENTS

PHASES OF THE MOON

	Full Moon	New Moon
2010	30 January	15 January
2011	19 January	4 January
2012	9 January	23 January
2013	27 January	11 January
2014	16 January	1, 30 January
2015	5 January	20 January
2016	24 January	10 January

ECLIPSES

2010: 15 January An annular eclipse of the Sun is visible from eastern Africa, the Indian Ocean, southern India, and southeastern Asia. A partial solar eclipse is seen from southern and eastern Africa, the Middle East, and central, southern, and southeastern Asia.

2011: 4 January A partial eclipse of the Sun is visible from northern Africa, Europe, the Middle East, and central Asia.

THE PLANETS

2010: 27 January
Mercury is at greatest morning elongation, magnitude –0.1.

2010: 29 January
Mars is at opposition, magnitude –1.2. At midnight, it is visible to the south from northern latitudes and to the north from southern latitudes.

2011: 8 January
Venus is at greatest morning elongation, magnitude –4.4.

2011: 9 January
Mercury is at greatest morning elongation, magnitude –0.2

2014: 6 January
Jupiter is at opposition, magnitude –2.7. At midnight it is visible to the south from northern latitudes and to the north from southern latitudes.

2014: 31 January
Mercury is at greatest evening elongation, magnitude –0.5.

2015: 14 January
Mercury is at greatest evening elongation, magnitude –0.6.

2015: 5 January
Mercury and Venus appear three Moon widths apart in the southwestern evening sky.

2016: 9 January
Venus and Saturn appear one-sixth of a Moon's width apart in the southeastern morning sky.

Opposition and **elongation** are explained on *page 26*

NORTHERN LATITUDES

Orion stands proud above the southern horizon with his dogs to the left of his feet. The dogs are located by their bright stars, brilliant Sirius in Canis Major, and Procyon, the eighth-brightest star of all, in Canis Minor. At either side of Orion's head are the constellations Gemini and Taurus. Gemini is identified by the two bright stars Castor and Pollux.

The head of Taurus, the bull, is close to the lion pelt in Orion's hand. The Hyades star cluster, easily visible to the naked eye, locates the bull's face, and the bright star Aldebaran marks its eye. Aldebaran is a red giant, and its colouring is apparent to the naked eye.

Directly overhead and within the path of the Milky Way is the constellation of Auriga, found by its bright star, Capella. The Milky Way flows into Cassiopeia, whose stars make a recognizable "W" or "M" shape in the sky, depending on whether they are above or below the north celestial pole.

Cassiopeia and the other circumpolar constellations are above the northern horizon. Cassiopeia is to the left of Polaris, the Pole Star, and the two bears, Ursa Minor and Ursa Major, are to its right.

The Quadrantid meteor shower occurs during the first week of January and is best seen after midnight. The meteors are usually faint and radiate from a point within the constellation Boötes, close to where Boötes borders the tail of Ursa Major. The short-lived peak of about 100 meteors an hour occurs on 3–4 January.

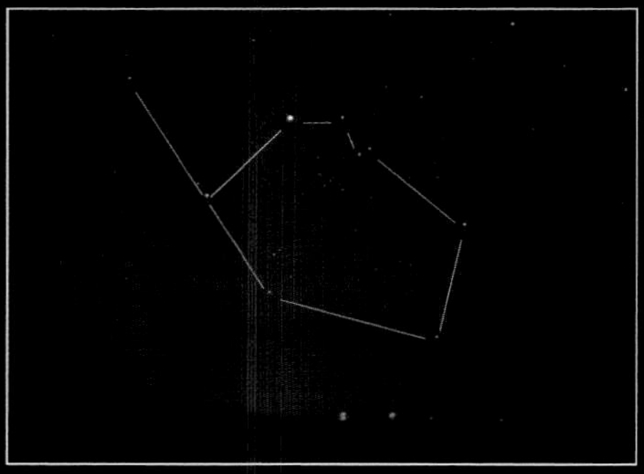

Capella and Auriga
Bright Capella (top centre) identifies the location of the constellation Auriga, the charioteer. It actually consists of two giant stars and has a yellow tint. It contains two star clusters, M36 and M38, just visible to the naked eye. Binoculars reveal them as fuzzy patches and confirm their presence.

MIDNIGHT
3AM
6AM
9AM
NOON
10°
0°
–10°
–20°
–30°
–40°
–50°
Arcturus
LEO
Regulus
CANCER
VIRGO
Spica
OPHIUCHUS
LIBRA
Antares
SAGITTARIUS
SCORPIUS
Shaula
MORNING SKY

Sirius and M41
Sirius (top centre) is the brightest star of all. Its brightness (magnitude –1.44) can be explained in part by its relative closeness; it is just 8.6 light years away. By contrast, the open star cluster M41 (bottom centre) is about 2,300 light years away. In good conditions, it can be seen with the naked eye.

SOUTHERN LATITUDES

Orion is high in the sky, and his head points down to the northern horizon while his feet are tipped towards the southern horizon. The three stars that form his belt are overhead. The sword hanging from the belt is identified by stars and a bright nebula, the Orion Nebula (M42), which is the brightest and best-known nebula. Sharp-eyed observers will see it as a milky patch, and binoculars will confirm its position. Close to M42 but nearer the belt is NGC 1981 – a scattered open star cluster that must be viewed through binoculars.

Auriga and its bright star Capella lie between Orion and the northern horizon. Procyon, the bright star of Canis Minor, is to the east of Betelgeuse, the star that marks one of the hunter's shoulders. To the east of Orion's feet is Sirius, the bright star that marks the head of the second of Orion's dogs. To the northwest and below Orion is Taurus; and to the lower right of Orion, beyond his club-holding hand, are the feet of Gemini, the twins. Closer still to the horizon, are the bright stars Castor and Pollux that mark the twins' heads.

The view looking south is one of contrasts. The eastern – that is, left – part of the sky is full of things to see, but just one bright star shines out in the western portion. This is Achernar, the ninth-brightest star. It marks one end of the meandering constellation Eridanus, which represents the mythological river of that name.

At the left, the Milky Way flows from the southeastern horizon and up into the sky towards Sirius. En route it crosses the rich starfields of Centaurus, Crux, and Carina. Centaurus, the centaur, is lying with its back almost against the horizon. Between its upward-pointing legs is Crux, the smallest constellation of all.

EVENING SKY

NIGHT
9PM
6PM
3PM
NOON
Capella
Castor
GEMINI
TAURUS
PLEIADES
ARIES
Aldebaran
HYADES
Bellatrix
Betelgeuse
Procyon
PISCES
Mira
Rigel
AQUARIUS
Fomalhaut
CAPRICORNUS
10°
0°
–10°
–20°
–30°
–40°
–50°

POSITIONS OF THE PLANETS

This chart shows the positions of Mercury, Venus, Mars, Jupiter, and Saturn in January from 2010 to 2016. The planets are represented by coloured dots, while the number inside the dot indicates the year. For all planets apart from Mercury, the dot indicates the planet's position on 15 January. Mercury is shown only when it is at greatest elongation (see p.26) – for the specific date of elongation, refer to the table on the facing page.

Mercury Venus Mars Jupiter Saturn

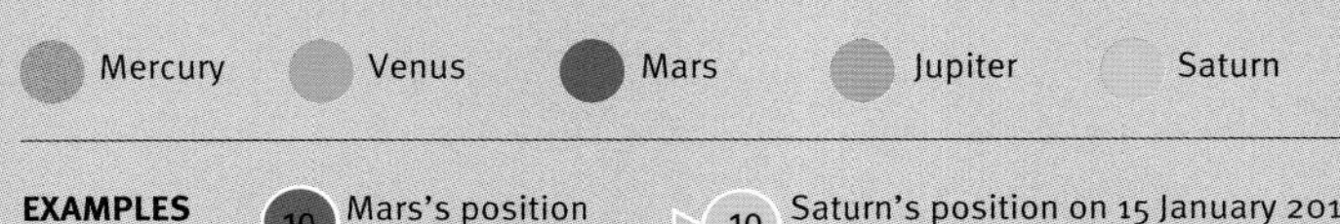

EXAMPLES 10 Mars's position on 15 January 2010. 10 Saturn's position on 15 January 2010. The arrow indicates that the planet is in retrograde motion (see p.26)

Orion the hunter
One of the most easily recognized constellations in the sky is Orion. Its stars effortlessly draw out the figure of the hunter. Single bright stars mark his shoulders, a small group of stars his head, and a line of three stars his belt. His sword is represented by stars and the star-forming region M42.

February

A triangle of bright stars and a pair of twins shine high in February's sky. The equilateral triangle is made by linking three stars in different constellations. They are Betelgeuse in Orion's shoulder, and Sirius and Procyon, the bright stars in each of Orion's two dogs, Canis Major and Canis Minor. The twins Castor and Pollux are represented by the bright stars of the same name.

SPECIAL EVENTS

PHASES OF THE MOON

	Full Moon	New Moon
2010	28 February	14 February
2011	18 February	3 February
2012	7 February	21 February
2013	25 February	10 February
2014	14 February	None
2015	3 February	18 February
2016	22 February	8 February

THE PLANETS

2013: 8 February
Mercury and Mars appear half a Moon's width apart in the western evening sky.

2013: 16 February
Mercury is at greatest evening elongation, magnitude –0.5.

2015: 7 February
Jupiter is at opposition, magnitude –2.6. At midnight it is visible to the south from northern latitudes and to the north from southern latitudes.

2015: 21 February
Venus and Mars appear a Moon width apart in the southwestern evening sky.

2015: 24 February
Mercury is at greatest morning elongation, magnitude 0.1.

2016: 7 February
Mercury is at greatest morning elongation, magnitude 0.0.

Opposition and **elongation** are explained on *page 26*

Beehive Cluster
Also known as Praesepe, or M44, the Beehive Cluster is in Cancer, seen high to the south this month. The brightest stars in this open cluster of 350 or so stars are magnitude 6, making it visible to the naked eye as a cloudy patch. Binoculars show it as a star field of more than three apparent Moon widths.

NORTHERN LATITUDES

The triangle of stars formed by Betelgeuse, Sirius, and Procyon is known as the Winter Triangle (see pp.32–33). Sirius and Procyon are easy to spot to the south, since they are the first- and eighth-brightest stars. Brilliant Sirius is closest to the southern horizon; Procyon is above, to Sirius's left. Betelgeuse is in the constellation of Orion, above and to the right of Sirius.

Beyond Orion, and in the southwestern sky, is Taurus, the bull. The distinctive shape of Leo, the lion, is now in view in the southeast, and Gemini is almost overhead. The twins are nearly parallel to the horizon, with their heads pointing to Leo and their feet towards Taurus.

Capella, the bright star in Auriga, is high in the western sky. Below is Perseus, the mythological hero who rescued Andromeda from a sea monster. The chained figure of Andromeda is between Perseus and the horizon. Her mother, Cassiopeia, is to her right in the northwestern sky. The two other circumpolar constellations, Ursa Major and Ursa Minor, are to the northeast.

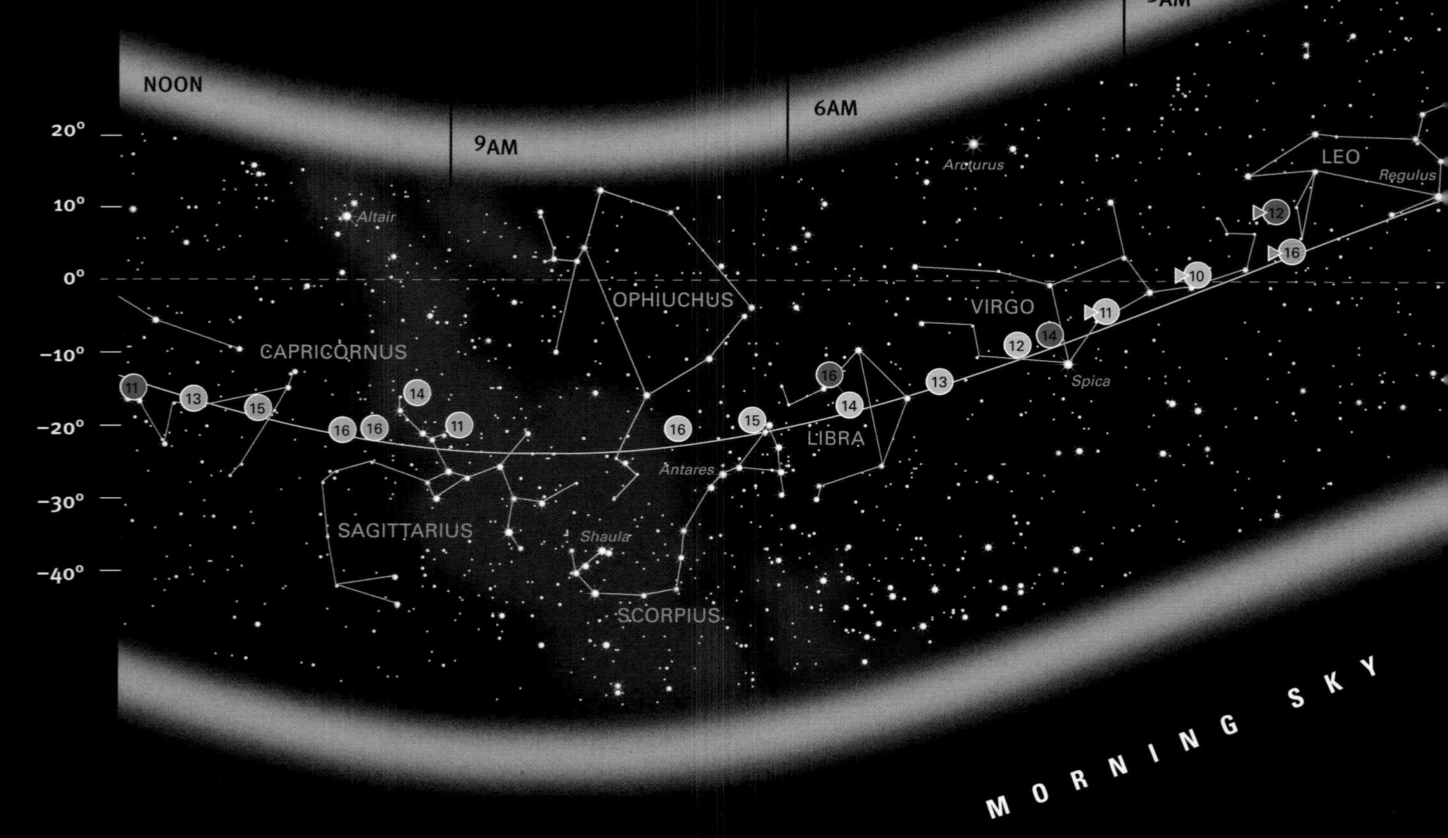

Rosette Nebula and NGC 2244
This spectacular telescope view of the Rosette Nebula in the constellation of Monoceros, the unicorn, reveals an open cluster of stars, NGC 2244, in the heart of the nebula. The surrounding nebula shows up well in photographs but not to the eye.

SOUTHERN LATITUDES

The path of the Milky Way extends across the sky, stretching from the southeastern horizon to the northwest. Along its path, starting closest to the southern horizon, are the constellations Centaurus, Crux, Carina, Vela, and Puppis, and then Canis Major, which is overhead. Its bright star, Sirius, and brilliant Canopus of Carina, are the two brightest stars of all. They are seen high in the sky throughout this month.

The equilateral triangle known to northern-hemisphere observers as the Winter Triangle is made by linking orange-red Betelgeuse and the white stars Sirius and Procyon. It is seen high above the northwestern horizon. Sirius is the highest star; below it are Betelgeuse and Procyon, to the left and right respectively. Within the triangle is the constellation Monoceros, the unicorn, which also lies in the path of the Milky Way. Close to its border with Orion is the open star cluster NGC 2244, which is visible through binoculars.

The stars Castor and Pollux, which mark the heads of the Gemini twins, are seen by looking north. At magnitude 1.2, Pollux is the brighter of the two, but Castor (magnitude 1.6) is possibly the more interesting. A small telescope will show it as a pair of stars. At right, to the east, is Cancer, the crab, and the upside-down figure of Leo, the lion. His head points toward the twins, his tail to the eastern horizon.

The Winter Triangle
Despite urban light pollution from the horizon, the Winter Triangle is visible here. Brilliant Sirius in Canis Major is just below the centre of the image. Above and to the left is Procyon in Canis Minor. At right centre is Orion; its star Betelgeuse forms the third point of the triangle.

EVENING SKY

POSITIONS OF THE PLANETS

This chart shows the positions of Mercury, Venus, Mars, Jupiter, and Saturn in February from 2010 to 2016. The planets are represented by coloured dots, while the number inside the dot indicates the year. For all planets apart from Mercury, the dot indicates the planet's position on 15 February. Mercury is shown only when it is at greatest elongation (see p.26) – for the specific date of elongation, refer to the table on the facing page.

Mercury · Venus · Mars · Jupiter · Saturn

EXAMPLES — Mars's position on 15 February 2010 · Saturn's position on 15 February 2010. The arrow indicates that the planet is in retrograde motion (see p.26)

March

Leo and Virgo take the place of Orion and Gemini as we herald the start of northern spring and southern autumn. On the 20th, or occasionally the 21st, of the month, the Sun crosses the celestial equator as it moves from the southern to the northern celestial sky. Briefly, day and night are of equal length before the northern-hemisphere nights grow shorter and the southern ones longer.

SPECIAL EVENTS

PHASES OF THE MOON

	Full Moon	New Moon
2010	30 March	15 March
2011	19 March	4 March
2012	8 March	22 March
2013	27 March	11 March
2014	16 March	1 March
2015	5 March	20 March
2016	23 March	9 March

ECLIPSES

2015: 20 March A total solar eclipse is visible from the Faroes (between Scotland and Iceland), the Norwegian Sea, and Svalbard.

2016: 9 March A total eclipse of the Sun is visible from Indonesia and the North Pacific.

THE PLANETS

2010: 22 March
Saturn is at opposition, magnitude 0.5. At midnight, it is visible to the south from northern latitudes and to the north from southern latitudes.

2011: 23 March
Mercury is at greatest evening elongation, magnitude 0.0.

2012: 3 March
Mars is at opposition, magnitude –1.2. At midnight, it is visible to the south from northern latitudes and to the north from southern latitudes.

2012: 5 March
Mercury is at greatest evening elongation, magnitude –0.3.

2012: 27 March
Venus is at greatest evening elongation, magnitude –4.3.

2013: 31 March
Mercury is at greatest morning elongation, magnitude 0.3.

2014: 14 March
Mercury is at greatest morning elongation, magnitude 0.2.

2016: March 8
Jupiter is at opposition, magnitude –2.5. At midnight it is visible to the south from northern latitudes and to the north from southern latitudes.

Opposition and **elongation** are explained on *page 26*

NORTHERN LATITUDES

Due south and high in the sky is the distinctive constellation Leo. Its linked stars really do resemble a crouching lion. The curve of his head and the shape of his body are easy to pick out in the sky. Leo looks towards the west, past the faint constellation of Cancer, and on to the winter constellation of Orion as it disappears from view. To the left of Leo are Virgo and its bright star Spica, rising over the eastern horizon. Leo's brightest star, Regulus, marks the start of an outstretched front leg. It is a blue-white star of magnitude 1.4. Through binoculars, a dimmer, wide companion can be seen. Just below the lion's body are five galaxies, all visible with binoculars in good observing conditions.

Ursa Major is situated high in the sky above the northern horizon. The bowl of the saucepan-shaped Plough, formed from seven bright stars in the hindquarters and tail of Ursa Major, is open downwards towards Polaris, the Pole Star. Among the stars near the bear's head is M81, one of the easiest galaxies to find with binoculars. It is a spiral galaxy also known as Bode's Galaxy. About one apparent Moon width away is a second, smaller and fainter galaxy, M82. Cassiopeia, on the opposite side of Polaris, is now almost between Polaris and the horizon, forming a "W" shape.

Galaxies in Ursa Major
Galaxies M81 and M82 are in Ursa Major, above the great bear's shoulder. With binoculars, the spiral galaxy M81 (right) appears as an elongated patch of light. Cigar-shaped M82 (left) is only visible through a telescope.

SOUTHERN LATITUDES

The view south is packed with stars. Due south are the constellations Carina and Vela. The white supergiant Canopus marks the western end of Carina. At magnitude −0.62, Canopus is second only to Sirius, the brightest star of all. Sirius is above and west of Canopus. Eta (η) Carinae is a variable star that brightens considerably when it erupts. In 1841, it rivalled Sirius after a giant outburst of material.

The southeastern sky is dominated by the two bright stars Alpha (α) and Beta (β) Centauri, which point to the nearby tiny constellation of Crux. Further east, the first of the winter constellations are rising above the eastern horizon. Virgo is leading the way, followed by Scorpius.

Due north is Leo. Its stars draw out a crouching lion, which southern-hemisphere observers see upside-down. The lion is apparently lying on its back with its head facing west; its bright star Regulus, marking the start of an outstretched front paw, is above. Almost equidistant and at either side of Regulus are two equally bright stars: to the west, Procyon in Canis Minor; and to the east, Spica in Virgo. Closer to the eastern horizon is the red giant Arcturus and the constellation Boötes, a forerunner of the winter sky to come. On the opposite side of the sky, the summer constellation of Orion can be seen disappearing over the western horizon.

Eta Carinae Nebula
The fuzzy red object at right is the Eta Carinae Nebula (NGC 3372). Embedded within is the variable star, Eta Carinae. To its left is Crux and the dark nebula the Coalsack, which is blocking the light of more distant stars.

Leo, the lion
The recognizable figure Leo is seen in this northern-sky view. The bright star at lower right is Regulus. The curve of stars that make up his shoulders and head extend up from Regulus. Denebola, the bright star at lower left, marks the start of the lion's tail. Turn the book upside-down to see how Leo appears in the southern sky.

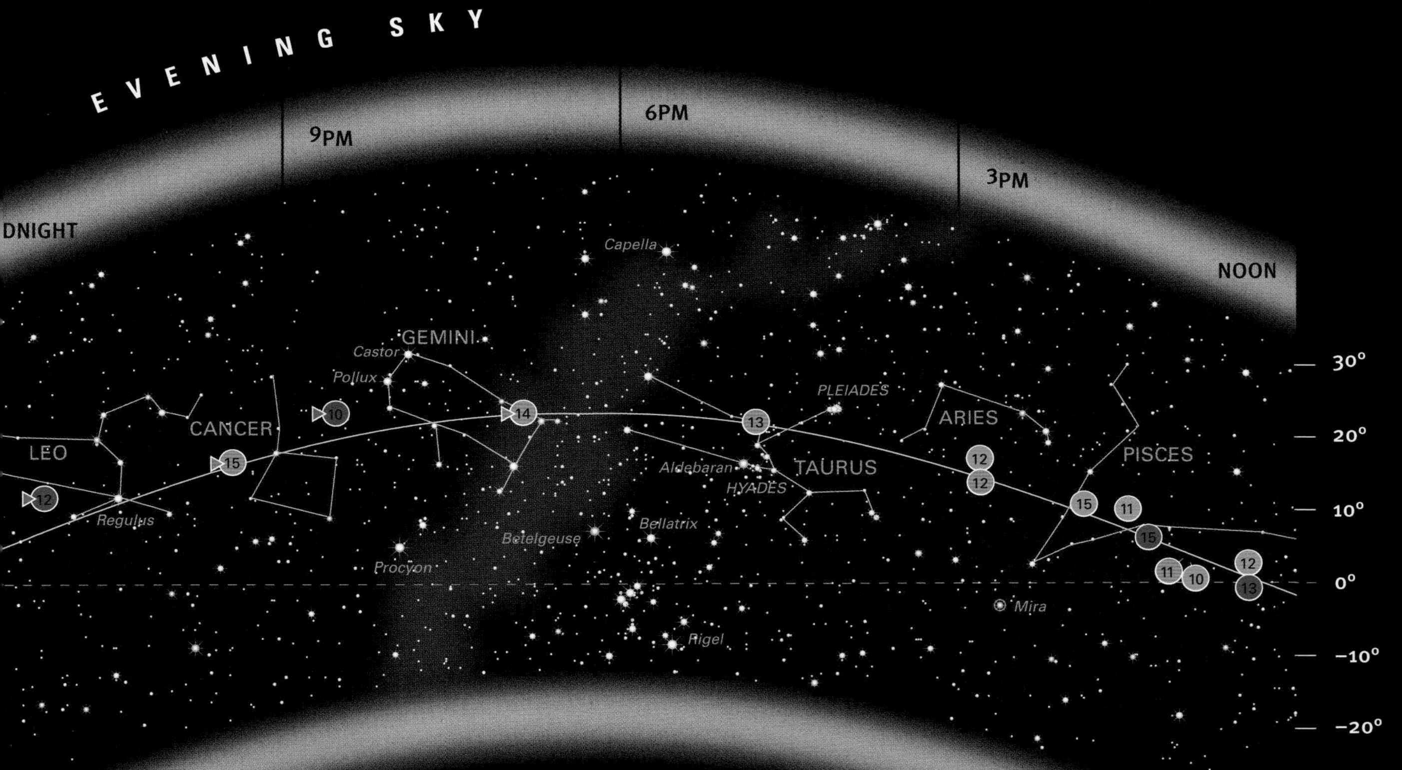

POSITIONS OF THE PLANETS

This chart shows the positions of Mercury, Venus, Mars, Jupiter, and Saturn in March from 2010 to 2016. The planets are represented by coloured dots, while the number inside the dot indicates the year. For all planets apart from Mercury, the dot indicates the planet's position on 15 March. Mercury is shown only when it is at greatest elongation (see p.26) – for the specific date of elongation, refer to the table on the facing page.

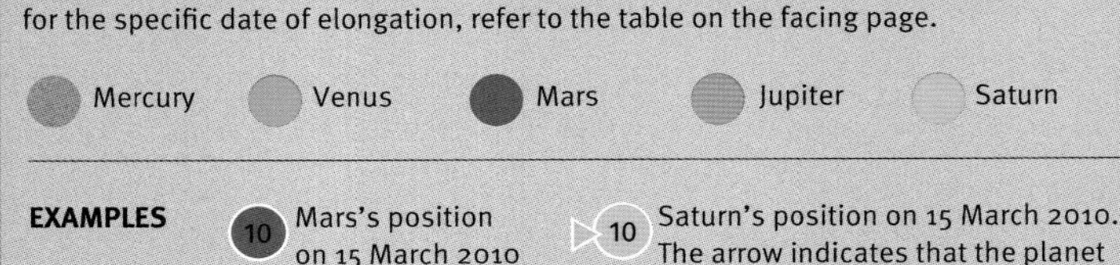

April

Leo remains high in the sky but now starts to make room for another zodiac constellation, Virgo. And a further mythical figure is making his presence felt. Boötes, the herdsman, is well placed for all observers. The third-largest constellation, Ursa Major, the great bear, is almost overhead for those in the northern hemisphere, while those observing from Earth's south find Crux prominently placed.

SPECIAL EVENTS

PHASES OF THE MOON

	Full Moon	New Moon
2010	20 April	14 April
2011	18 April	3 April
2012	6 April	21 April
2013	25 April	10 April
2014	15 April	29 April
2015	4 April	18 April
2016	22 April	7 April

ECLIPSES

2014: 15 April A total eclipse of the Moon is visible from the Americas.

2014: 29 April An annular solar eclipse is visible in Antarctica and will be seen as partial over Australia.

2015: 4 April A brief total eclipse of the Moon is visible from east Asia, Australia, and western North America.

THE PLANETS

2010: 8 April
Mercury is at greatest evening elongation, magnitude 0.0.

2011: 4 April
Saturn is at opposition, magnitude 0.4. At midnight, it is visible to the south from northern latitudes and to the north from southern latitudes.

2012: 15 April
Saturn is at opposition, magnitude 0.2. At midnight, it is visible to the south from northern latitudes and to the north from southern latitudes.

2012: 18 April
Mercury is at greatest morning elongation, magnitude 0.5.

2013: 28 April
Saturn is at opposition, magnitude 0.13. At midnight it is visible to the south from northern latitudes and to the north from southern latitudes.

2014: 8 April
Mars is at opposition, magnitude –1.5. At midnight it is visible to the south from northern latitudes and to the north from southern latitudes.

2016: 18 April
Mercury is at greatest evening elongation, magnitude 0.2.

Opposition and **elongation** are explained on *page 26*

NORTHERN LATITUDES

Leo is high in the southwest sky, its distinctive shape easily picked out. The lion is looking towards the heads of the twins, Gemini, marked by the bright stars Castor and Pollux. The sky above and below is relatively barren. The long figure of Hydra, the water snake, straggles between Leo and the southern horizon. This is the largest constellation, but it is far from prominent. To the right, Virgo follows Leo across the sky. Its star Spica shines bright in the southeast. Above and to the east is Arcturus, the brightest star north of the celestial equator. At magnitude –0.05, this red giant is the fourth-brightest star.

Looking north, Ursa Major is directly above Polaris, the Pole Star, and almost overhead. Seven stars that form the tail and rump of the bear make a saucepan shape in profile. Known as the Plough, or Big Dipper, this is one of the most recognized star patterns in the northern hemisphere. Cassiopeia is below, close to the horizon. To the northwest is the yellow-coloured Capella, which, at magnitude 0.08, is the sixth-brightest star. It is the brightest and by far the most conspicuous star in Auriga, the charioteer. To the east, Vega in Lyra heralds the arrival of the first of the summer constellations. Lyra is host to the Lyrids meteor shower, which peaks around 21–22 April. A dozen or so meteors can be seen every hour radiating from a point near Vega.

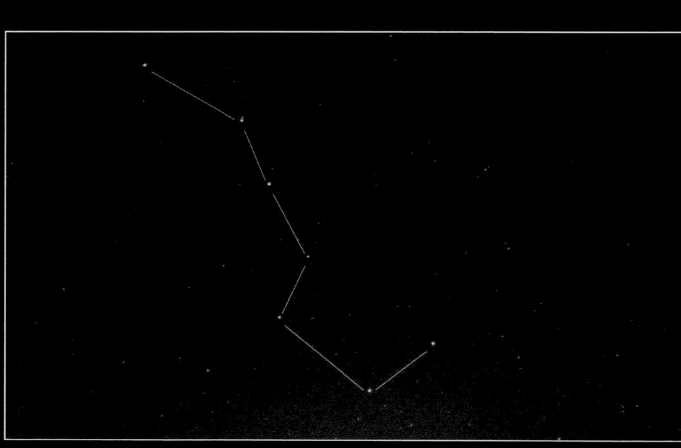

Ursa Major and the Plough
The seven stars in the great bear's tail and rump (top and centre) form the Plough. The second star in the tail is Mizar, which has a companion, Alcor. Both can be seen with binoculars or by anyone with good eyesight.

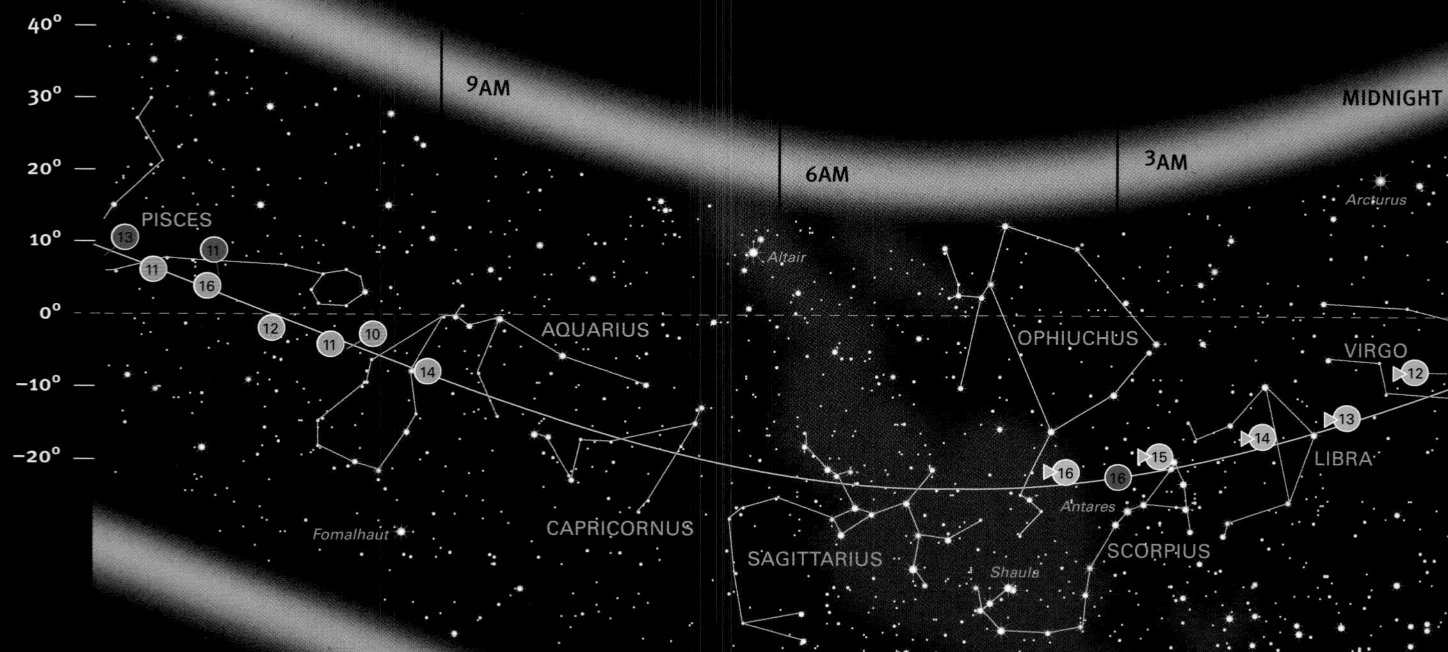

MORNING SKY

Crux and Carina
The Milky Way flows from top to bottom in this view of the southern-hemisphere sky. The four stars that form the southern cross, Crux, can be seen at the top left of the image. Below are the stars of Carina, which represent the keel of a ship.

SOUTHERN LATITUDES

The star-rich path of the Milky Way makes a glorious sight in the southern sky. It extends from the western horizon and climbs high in the sky to the south, before descending to the east. En route it passes Sirius and incorporates Carina, Crux and Centaurus, and the tail of Scorpius, the scorpion. Crux is almost due south and well placed for observation. The sky overhead, above Crux, is barren by comparison. It contains the long meandering body of Hydra, the water snake.

Acux, the brightest star in the constellation Crux, marks the base of the cross. To the naked eye it is a single star of magnitude 0.8, but through a small telescope it is seen as a pair of stars. Beta (β) Crucis marks the left of the cross; it is a blue-white giant of magnitude 1.3. Between this star and the prominent dark nebula known as the Coalsack is the sparkling Jewel Box cluster. Also known as NGC 4755, this is an open star cluster visible to the naked eye. Binoculars or a small telescope will show its individual stars.

Looking north, Leo is still well placed but is heading for the western horizon. To the right of its bright star Regulus are five galaxies visible with binoculars. The bright stars Spica in Virgo and Arcturus in Boötes shine bright in the northeastern sky. Procyon is still visible in the northwest, but Gemini is now moving below the horizon.

The Sombrero Galaxy
Also known as M104 or NGC 4594, the Sombrero Galaxy is a spiral galaxy in Virgo. Its dark dust lane and bulbous core resemble the traditional Mexican hat after which it is named. It is seen almost edge-on and appears elongated in a small telescope. The dark lane is visible through a larger instrument.

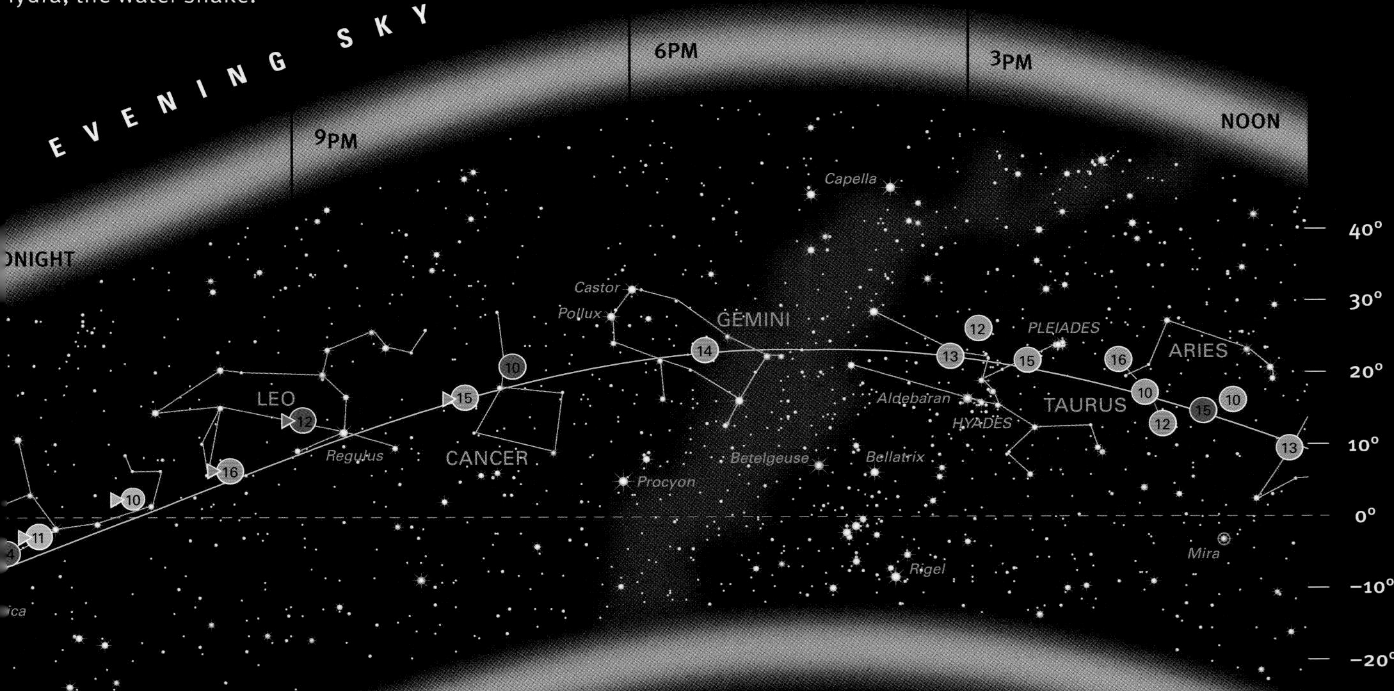

POSITIONS OF THE PLANETS

This chart shows the positions of Mercury, Venus, Mars, Jupiter, and Saturn in April from 2010 to 2016. The planets are represented by coloured dots, while the number inside the dot indicates the year. For all planets apart from Mercury, the dot indicates the planet's position on 15 April. Mercury is shown only when it is at greatest elongation (see p.26) – for the specific date of elongation, refer to the table on the facing page.

Mercury · Venus · Mars · Jupiter · Saturn

EXAMPLES — 10 Mars's position on 15 April 2010 — 10 Saturn's position on 15 April 2010. The arrow indicates that the planet is in retrograde motion (see p.26)

May

Boötes and Virgo are prominent for May observers in both the northern and southern hemispheres. Those to the south of the equator also see the constellations Crux and Centaurus, which are now at their highest above the horizon. In the northern hemisphere, the days are lengthening as summer approaches. Once the sky has darkened, Ursa Major stands high and proud in the northern sky.

SPECIAL EVENTS

PHASES OF THE MOON

	Full Moon	New Moon
2010	27 May	14 May
2011	17 May	3 May
2012	6 May	20 May
2013	25 May	10 May
2014	14 May	28 May
2015	4 May	18 May
2016	21 May	6 May

ECLIPSES

2012: 20–21 May An annular eclipse of the Sun is visible from the northern Pacific Ocean, southern Japan, and the western United States. A partial solar eclipse is seen from northeast Asia, the northern Pacific Ocean, and western North America.

2013: 10 May An annular eclipse of the Sun is visible from northern Australia and into the central Pacific Ocean.

THE PLANETS

2010: 26 May
Mercury is at greatest morning elongation, magnitude 0.5.

2011: 1 May
Mars and Jupiter appear about a Moon width apart in the eastern dawn sky.

2011: 7 May
Mercury is at greatest morning elongation, magnitude 0.5.

2011: 7–20 May
Mercury and Venus appear three Moon widths apart in the eastern dawn sky.

2011: 11 May
Venus and Jupiter appear about a Moon width apart in the eastern dawn sky.

2011: 21 May
Mercury and Mars appear about four Moon widths apart in the eastern dawn sky.

2011: 23–24 May
Venus and Mars appear two Moon widths apart in the eastern dawn sky.

2013: 25–28 May
Mercury, Venus, and Jupiter appear within five Moon widths of each other in the western evening sky.

2014: 10 May
Saturn is at opposition, magnitude 0.06. At midnight it is visible to the south from northern latitudes and to the north from southern latitudes.

2014: 25 May
Mercury is at greatest evening elongation, magnitude 0.6.

2015: 7 May
Mercury is at greatest evening elongation, magnitude 0.5.

2015: 23 May
Saturn is at opposition, magnitude 0.02. At midnight it is visible to the south from northern latitudes and to the north from southern latitudes.

2016: 22 May
Mars is at opposition, magnitude –2.1. At midnight it is visible to the south from northern latitudes and to the north from southern latitudes.

Opposition and **elongation** are explained on *page 26*

NORTHERN LATITUDES

The bright star Spica and its constellation, Virgo, are due south. Higher in the sky, and almost overhead is Arcturus in Boötes. The 13th "zodiac" constellation, Ophiuchus, is seen to the southeast. The head and body of Leo are still visible in the southwest, before sinking below the horizon. Antares, the red star that marks the heart of Scorpius, the scorpion, makes a rare appearance in the northern sky for those observers south of 50°N. It can be seen close to the horizon for the summer months.

Ursa Major is high in the sky to the north. The easily recognized saucepan shape of the Plough is positioned so that the end of the handle is north and the pan tips down to the west. Dubhe and Merak, the two brightest stars in Ursa Major, form the side of the pan away from the handle. These are known as the pointers, since they point the way to Polaris. Also known as the Pole Star, Polaris marks the position of the north celestial pole.

The summer stars Vega (in Lyra) and Deneb (in Cygnus) move higher into the northeastern sky as the last of the winter stars, Castor and Pollux (in Gemini), set in the northwest. Lyra is small but prominent due to its bright star Vega. It is also home to the planetary nebula, the Ring Nebula (M57), seen as a misty disc through a small telescope. The Eta Aquarid meteor shower (see opposite) is visible to observers in the lower latitudes.

Mizar and Alcor
This image shows the stars Mizar (left) and Alcor (right). They are usually seen as one star in the handle of the Plough. When seen through a telescope (above), Mizar is found to have another companion.

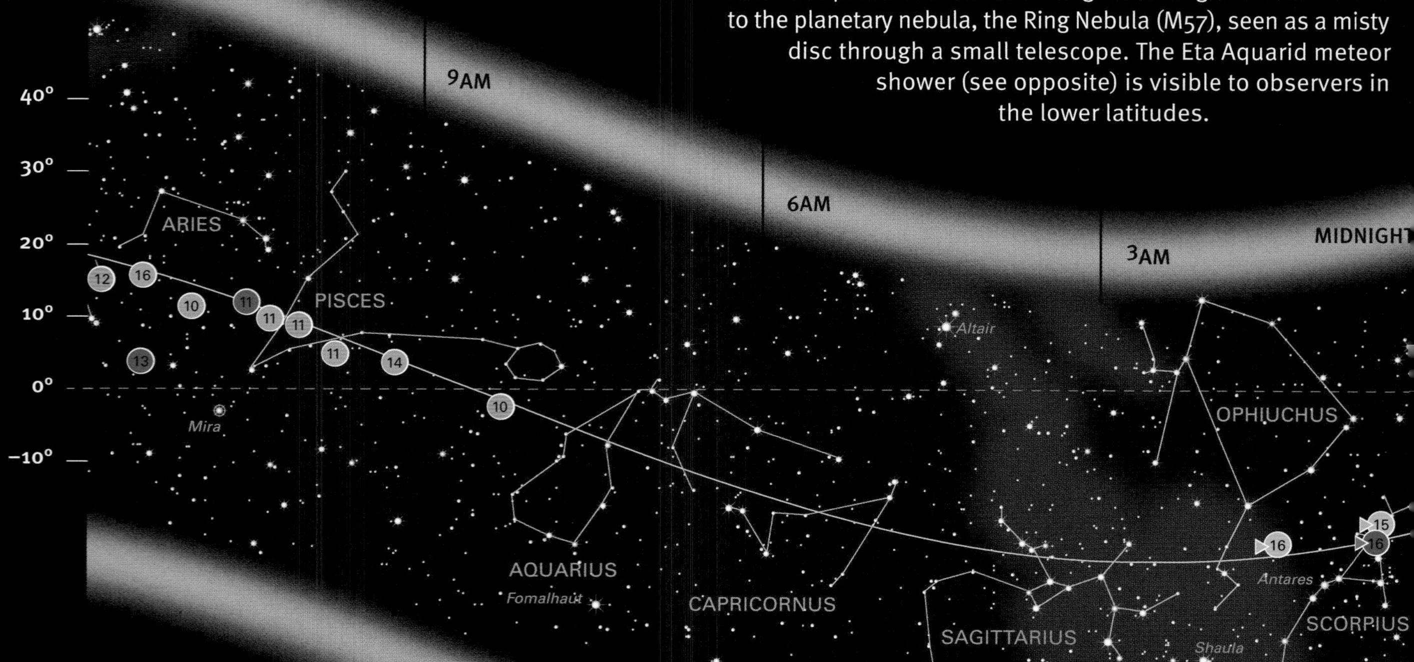

SOUTHERN LATITUDES

Two bright stars, Spica (in Virgo) and Arcturus (Boötes), are centrally placed to the south. Arcturus is the lower and brighter of the two and the fourth-brightest star of all. Spica is almost overhead. To the northwest is Leo, which will soon be making way for the winter constellations. Ophiuchus, the serpent holder, is in the eastern sky. The serpent, which forms the constellation Serpens, is coiled around Ophiuchus – it is in two halves, one at either side of Ophiuchus. The head, Serpens Caput, is between Ophiuchus and Boötes; the tail, Serpens Cauda, is close to the eastern horizon.

Centaurus and Crux are high in the sky to the south. The two bright stars of Centaurus shine out, even against the starry path of the Milky Way. Alpha (α) Centauri is the brighter of the two and third brightest of all. Within the centaur's body is Omega (ω) Centauri, the brightest globular cluster in all of Earth's sky. This month also offers the first good opportunity to look at Scorpius, a winter constellation that is now rising in the southeastern sky.

The Eta Aquarid meteor shower starts in late April but peaks in the first week of May, when about 35 meteors an hour can be seen. The meteors radiate from a point in Aquarius. They are visible in the hours before dawn as Earth ploughs into a stream of dust lost by Halley's Comet.

Alpha Centauri
By the horizon in this dawn sky, Alpha Centauri, just 4.3 light years away, is often said to be our closest star after the Sun. An unseen companion, Proxima Centauri, is in fact slightly closer and truly the closest night-time star.

Omega Centauri
The largest globular cluster in the Milky Way is Omega Centauri (NGC 5139). Just some of its 10 million or so stars are visible in this 4m (13ft) telescope image. To the naked eye or through binoculars, it appears as a fuzzy star.

Boötes
The large figure of Boötes, the herdsman, raises his arm high in the sky. The top star of the kite shape marks his head. At the tail end of the kite is the red giant Arcturus, the brightest star north of the celestial equator.

EVENING SKY

POSITIONS OF THE PLANETS

This chart shows the positions of Mercury, Venus, Mars, Jupiter, and Saturn in May from 2010 to 2016. The planets are represented by coloured dots, while the number inside the dot indicates the year. For all planets apart from Mercury, the dot indicates the planet's position on 15 May. Mercury is shown only when it is at greatest elongation (see p.26) – for the specific date of elongation, refer to the table on the facing page.

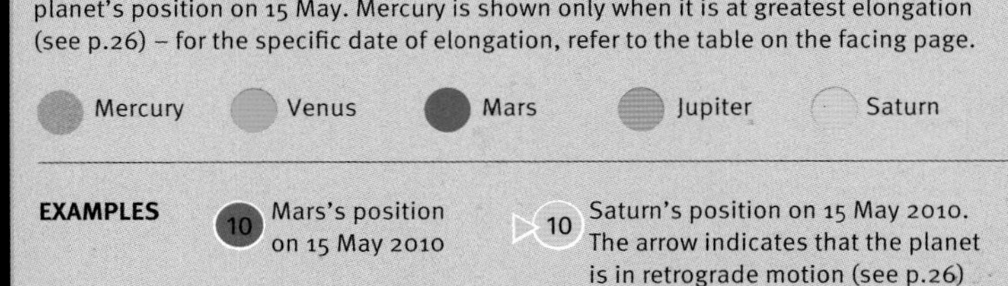

June

The mythical hero Hercules moves into the June sky to join Boötes, the herdsman, and Ophiuchus, the serpent holder. Towards the end of the month, the nights are at their shortest for northern observers and at their longest for those in the southern hemisphere. This is because on 21 June, the Sun is at its farthest point north of the celestial equator.

SPECIAL EVENTS

PHASES OF THE MOON

	Full Moon	New Moon
2010	26 June	12 June
2011	15 June	1 June
2012	4 June	19 June
2013	23 June	8 June
2014	13 June	27 June
2015	2 June	16 June
2016	20 June	5 June

ECLIPSES

2010: 26 June A partial eclipse of the Moon is visible from the Pacific Ocean, Australasia, and eastern Asia.

2011: 1 June A partial eclipse of the Sun is visible from Greenland, northern North America, and parts of northern and northeastern Asia.

2011: 15 June A total eclipse of the Moon is visible from Australia, southern Asia, the Indian Ocean, Africa, and Europe.

2012: 4 June A partial eclipse of the Moon is visible from western North and South America, the Pacific Ocean, Australasia, and eastern Asia.

THE PLANETS

2012: 5–6 June
A transit of Venus across the Sun is visible from North America, the Pacific Ocean, Australasia, and Asia.

2012: 30 June
Venus and Jupiter appear about ten Moon widths apart in the eastern dawn sky.

2013: 12 June
Mercury is at greatest evening elongation, magnitude 0.6.

2015: 24 June
Mercury is at greatest morning elongation, magnitude 0.6.

2016: 3 June
Saturn is at opposition, magnitude 0.0. At midnight it is visible to the south from northern latitudes and to the north from southern latitudes.

2016: 5 June
Mercury is at greatest morning elongation, magnitude 0.6.

Opposition and **elongation** are explained on *page 26*

NORTHERN LATITUDES

When looking south, Hercules and Boötes are high in the sky. The bright star Arcturus makes Boötes easy to find. Hercules, although large, is not prominent. It is best found by locating Vega in the east. Vega is in the constellation Lyra and is the fifth-brightest star of all. It is outshone in the June sky only by Arcturus. To the right of Vega is the body of Hercules. Four linked stars that form a distorted square, called the Keystone, represent his lower torso.

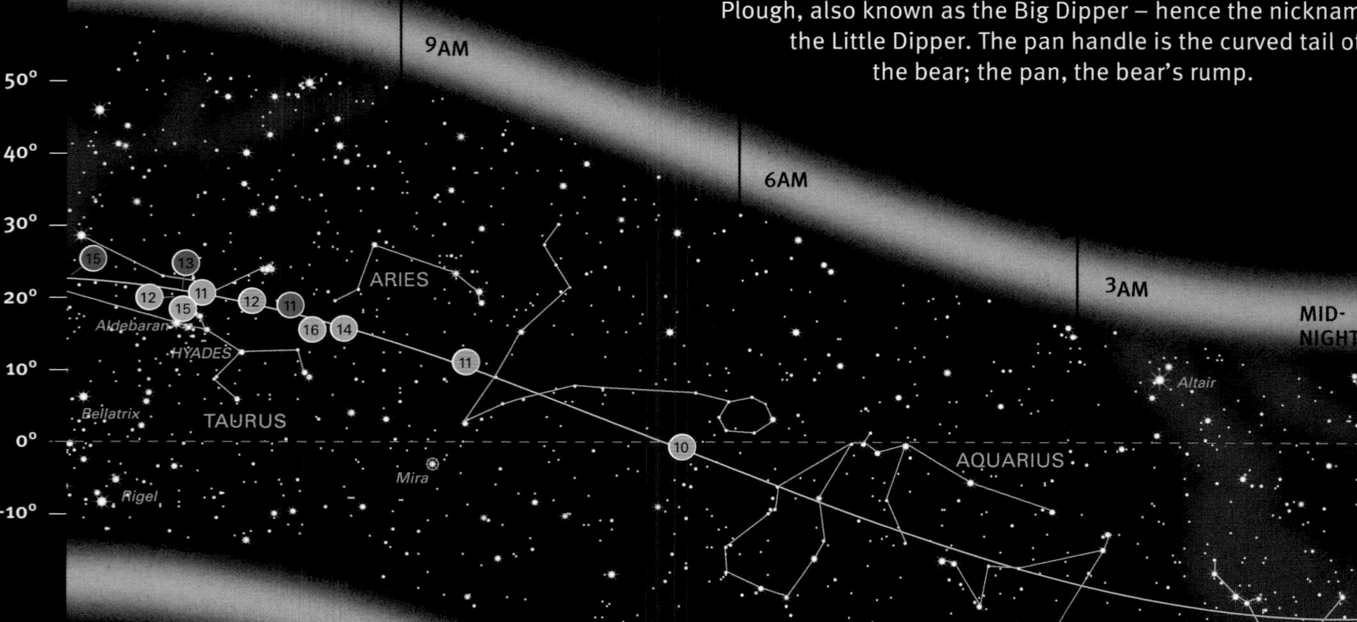

Albireo
The second brightest star in Cygnus, Albireo marks the swan's beak. It appears like a single star with the naked eye, but when viewed with powerful binoculars or a telescope, it is seen as a double star.

His legs point upwards, and his head points towards the horizon. The star cluster M13 can be spotted by the naked eye and lies on a line linking two of the stars. The head of Scorpius is just above the horizon, located by spotting red supergiant Antares.

In the east are three stars that form the asterism known as the Summer Triangle. The brightest and highest is Vega, and nearer to the horizon, Deneb (Cygnus) and Altair (Aquila) can be seen. Due north, Ursa Minor extends above Polaris, the Pole Star, which marks the tip of the bear's tail. The stars in Ursa Minor have no direct resemblance to a bear; indeed, they take the form of a saucepan in profile. Their shape echoes that of the Ursa Major asterism the Plough, also known as the Big Dipper – hence the nickname the Little Dipper. The pan handle is the curved tail of the bear; the pan, the bear's rump.

MORNING SKY

SOUTHERN LATITUDES

Boötes and Hercules are either side of due north. The figures of the huntsman and hero are found by locating the bright stars Arcturus and Vega, which is just peeking above the northeastern horizon. On the left, Boötes is head down and feet high; at right, Hercules is head high and feet down, with an arm outstretched above Vega. His lower body is represented by four linked stars known as the Keystone. The star cluster M13 lies on a line linking two of the stars. Aquila and its bright star Altair are now in the northeast.

Overhead, the distinctive shape of Scorpius with its curving tail is clearly seen. Its heart is marked by Antares, a red supergiant whose brightness varies between magnitudes 0.9 and 1.2 over a four-to-five-year period. Two open-star clusters, M6 and M7, are found near the sting in the scorpion's tail. Both are visible to the naked eye.

Scorpius and its neighbour Sagittarius lie in the Milky Way and in the direction of the centre of our galaxy. From Sagittarius and Scorpius, the milky path of light flows to the southwest, taking in the constellations Centaurus and Crux along the way. The globular cluster Omega (ω) Centauri is well placed, as are the dark Coalsack nebula and the Jewel Box (NGC 4755) open cluster, both in Crux.

Butterfly Cluster
The Butterfly Cluster (M6) is an open cluster in Scorpius. It is about 2,000 light years away and about 12 light years across. Its brightest star is an orange giant whose light varies over time. The star cluster is found near the sting of the scorpion's tail.

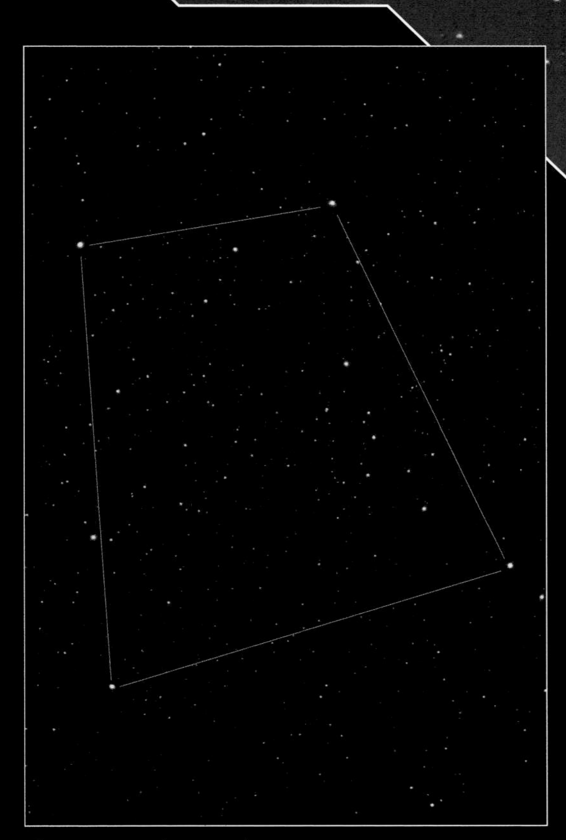

The Keystone in Hercules
The four stars that form the lower body of Hercules are known as the Keystone. The globular cluster M13 is bisected by a line joining two of the stars (seen here at left). The cluster can be seen from both northern and southern latitudes.

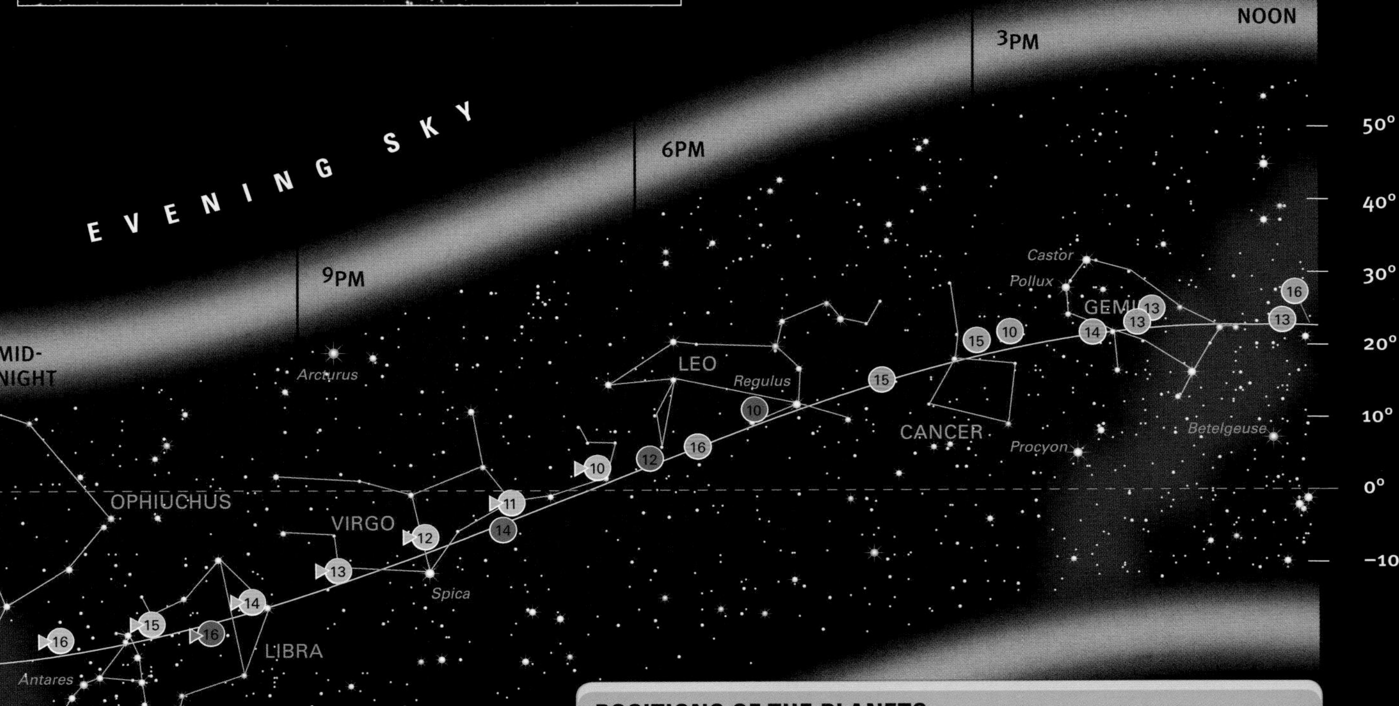

POSITIONS OF THE PLANETS

This chart shows the positions of Mercury, Venus, Mars, Jupiter, and Saturn in June from 2010 to 2016. The planets are represented by coloured dots, while the number inside the dot indicates the year. For all planets apart from Mercury, the dot indicates the planet's position on 15 June. Mercury is shown only when it is at greatest elongation (see p.26) – for the specific date of elongation, refer to the table on the facing page.

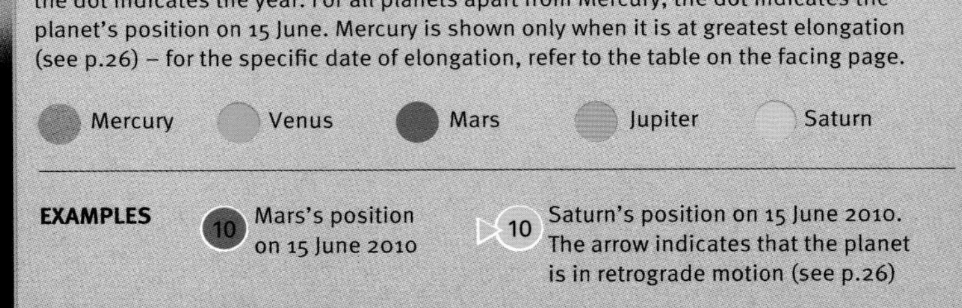

July

Hercules and Ophiuchus remain centre stage, and Aquila, the eagle, has flown into view. Southern observers are treated to the rich star fields of the Milky Way. These include Sagittarius, within which, almost directly overhead, lies the centre of our galaxy. The outstretched wings of Cygnus, the swan, are now high in the sky for northern observers. Vega, in the small constellation, Lyra, shines brightly overhead.

SPECIAL EVENTS

PHASES OF THE MOON

	Full Moon	New Moon
2010	26 July	11 July
2011	15 July	1, 30 July
2012	3 July	19 July
2013	22 July	8 July
2014	12 July	26 July
2015	2, 31 July	16 July
2016	20 July	4 July

ECLIPSES

2010: 11 July A total eclipse of the Sun is visible from the southern Pacific Ocean.

THE PLANETS

2010: 30–31 July
Mars and Saturn appear about four Moon widths apart in the western evening sky.

2011: 20 July Mercury is at greatest evening elongation, magnitude 0.5.

2012: 1 July Mercury is at greatest evening elongation, magnitude 0.5.

2012: 1–2 July Venus and Jupiter appear about ten Moon widths apart in the eastern dawn sky.

2013: 22 July
Mars and Jupiter appear two Moon widths apart in the morning eastern sky.

2013: 30 July
Mercury is at greatest morning elongation, magnitude 0.3.

2014: 12 July
Mercury is at greatest morning elongation, magnitude 0.3.

2015: 1 July
Venus and Jupiter appear one Moon width apart in the western evening sky.

Elongation is explained on *page 26*

NORTHERN LATITUDES

Looking north, Ursa Minor reaches up from Polaris. Coiling around it is the long, winding figure of Draco, the dragon. Ursa Major is to the left of Polaris, in the northwest; to the right, in the northeast, are the mythical king Cepheus and his queen, Cassiopeia. Arcturus (Boötes), in the west, and Vega (Lyra), almost overhead, are the brightest stars in the July sky.

Next in brightness is Altair in the southeast, marking the neck of Aquila, the eagle, and flanked by two dimmer stars. The Milky Way flows through Aquila and on to Cygnus, high in the east. A line of stars forms the swan's body and an intersecting line its wings; this cross shape led to its other name, the northern cross. Cygnus's brightest star, the blue-white supergiant Deneb, marks its tail. One of the smallest constellations, Corona Borealis is high in the southwest; it is flanked by Hercules and Boötes. Its arc of seven stars depicts the crown of the mythical Princess Ariadne, hence its alternative name, the northern crown. Ophiuchus is well placed due south, with Hercules above. Observers south of about 45°N will see star-rich Sagittarius and Scorpius on their horizons. This is the best chance for northern observers to see the two most southerly constellations of the zodiac in the evening sky.

Vega
The blue-white star Vega (below centre) is the brightest star in the northern summer sky. Close by is the star Epsilon (ε) Lyrae. Sharp eyesight or binoculars will show it as a double star (upper left). A telescope reveals that each of its two stars is also a double.

POSITIONS OF THE PLANETS

This chart shows the positions of Mercury, Venus, Mars, Jupiter, and Saturn in July from 2010 to 2016. The planets are represented by coloured dots, while the number inside the dot indicates the year. For all planets apart from Mercury, the dot indicates the planet's position on 15 July. Mercury is shown only when it is at greatest elongation (see p.26) – for the specific date of elongation, refer to the table to the left.

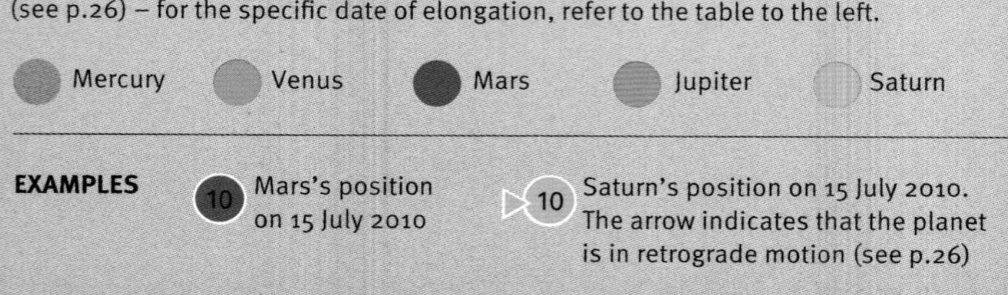

The Lagoon Nebula
The bright, star-forming Lagoon Nebula (M8, NGC6523) is some 5,200 light years away, within the archer's bow of the constellation of Sagittarius. It is visible to the naked eye in a rural sky and is a good binocular object. Binoculars will reveal the cluster of stars, NGC 6530, within the nebula.

SOUTHERN LATITUDES

Scorpius and Sagittarius are almost overhead. The scorpion's tail is due south, and so the open clusters M6 and M7 are ideally placed for observation. The red supergiant Antares shines out. It is the brightest star in Scorpius and one of the largest stars visible to the naked eye. Estimates of its size range from 280 to 700 times that of the Sun.

The view towards Sagittarius is towards the centre of our galaxy. A large and bright field of stars, M24 in Sagittarius, is visible to the naked eye; so, too, is the Lagoon Nebula (M8). M22, the third-brightest globular cluster, is also in Sagittarius. Easy to find with binoculars, it can also be seen by keen-eyed observers in a rural sky. A group of eight stars within Sagittarius forms the Teapot asterism, but it is difficult to pick out in this star-studded part of the sky. The arc of stars at the forefeet of Sagittarius is easier to see. It is Corona Australis, the southern crown, and one of the smallest constellations of all.

Looking north, Ophiuchus is high in the sky; below it is the figure of Hercules. Vega in Lyra is close to the horizon. Adjoining Vega is the northern-hemisphere constellation Cygnus. Observers to the north of about 30°S are able to see the complete figure of the flying swan. Above it in the northeastern sky is Aquila, the eagle.

The Delta Aquarid meteor shower reaches its peak around the 29th of this month. Up to 20 rather faint meteors an hour radiate from the southern half of Aquarius.

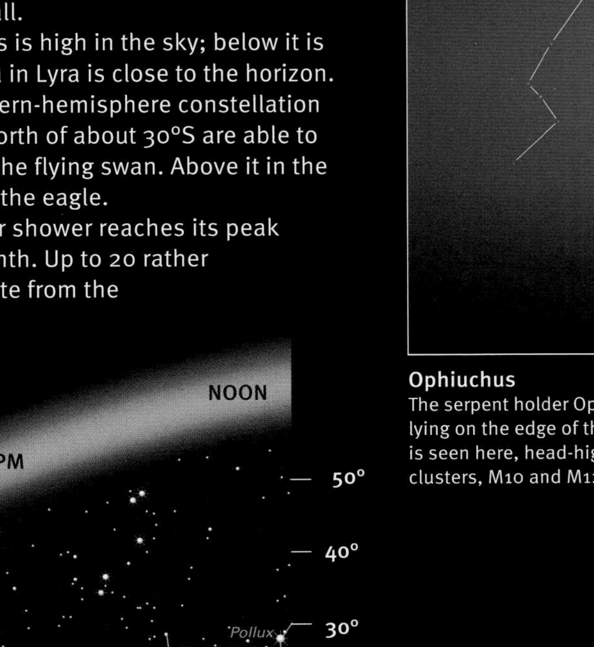

Ophiuchus
The serpent holder Ophiuchus is the 11th-largest constellation, lying on the edge of the Milky Way's starry path. The figure is seen here, head-high. Within his body are two globular clusters, M10 and M12 – both are visible through binoculars.

August

The Summer Triangle, a triangle of three brilliant stars from three different constellations, is seen from all but the most southerly latitudes. Northern observers have a glimpse towards the galaxy's centre and are treated to the Perseids, the best meteor shower of the year. Sagittarius remains well-placed for southern observers, and August also offers them the best chance of observing two celebrated planetary nebulae.

SPECIAL EVENTS

PHASES OF THE MOON

	Full Moon	New Moon
2010	24 August	10 August
2011	13 August	29 August
2012	2, 31 August	17 August
2013	21 August	6 August
2014	10 August	25 August
2015	29 August	14 August
2016	18 August	2 August

THE PLANETS

2010: 7 August
Mercury is at greatest evening elongation, magnitude 0.4.

2010: 20 August
Venus is at greatest evening elongation, magnitude –4.3.

2012: 15 August
Venus is at greatest morning elongation, magnitude –4.3.

2012: 16 August
Mercury is at greatest morning elongation, magnitude 0.3.

2014: 18 August
Venus and Jupiter appear half a Moon width apart in the eastern morning sky.

2014: 27 Augustt
Mars and Saturn appear seven Moon widths apart in the southwestern evening sky.

2016: 16 August
Mercury is at greatest evening elongation, magnitude 0.3.

2016: 25 August
Mars and Saturn appear nine Moon widths apart in the southern sky at sunset.

2016: 27 August
Venus and Jupiter appear one-sixth of a Moon width apart in the western evening sky with Mercury 10 Moon widths away.

Elongation is explained on *page 26*

The Summer Triangle
Three bright stars make up the Summer Triangle. Deneb in Cygnus, at top left, is the dimmest of the three; the brightest, Vega in Lyra, is to the right; and Altair in Aquila is seen at lower left.

NORTHERN LATITUDES

The stars of the Summer Triangle, a familiar feature in northern summer and autumn skies, are overhead. As the sky darkens, blue-white Vega in Lyra is the first triangle star to appear. To its east is Cygnus, the swan, which contains the second star, blue-white supergiant Deneb. The third, Altair in Aquila, the eagle, is below, to the south.

Sagittarius, the archer, is a centaur, a mythical beast. His human head and upper body are just above the southern horizon; his horse's legs are only visible to observers south of about 40°N. Sagittarius marks the centre of the galaxy. Here the Milky Way is at its densest and brightest, but a dark horizon is needed to reveal the star-studded path. It flows up from the horizon and although the path becomes dimmer, it may be easier to see. The path moves through Aquila, overhead into Cygnus, and on to Cassiopeia in the northeastern sky.

The best meteor shower of the year, the Perseids, peaks around 12 August, with up to 80 meteors an hour, radiating from a point in Perseus. Although Perseus is not clear of the eastern horizon before midnight, some pre-midnight meteors may be seen.

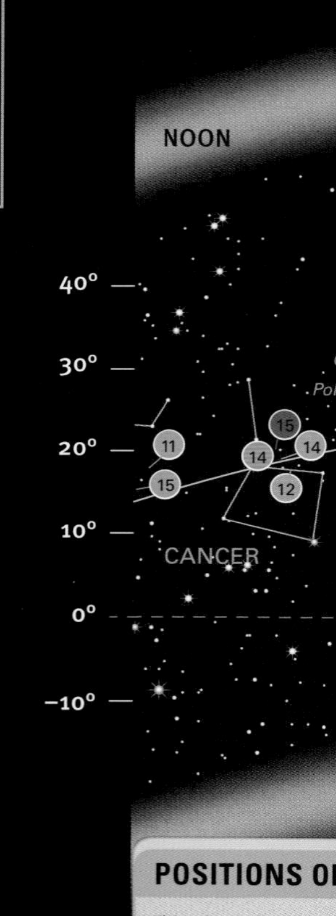

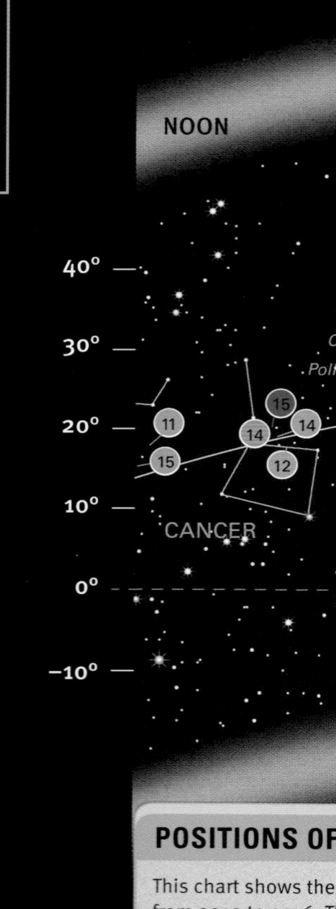

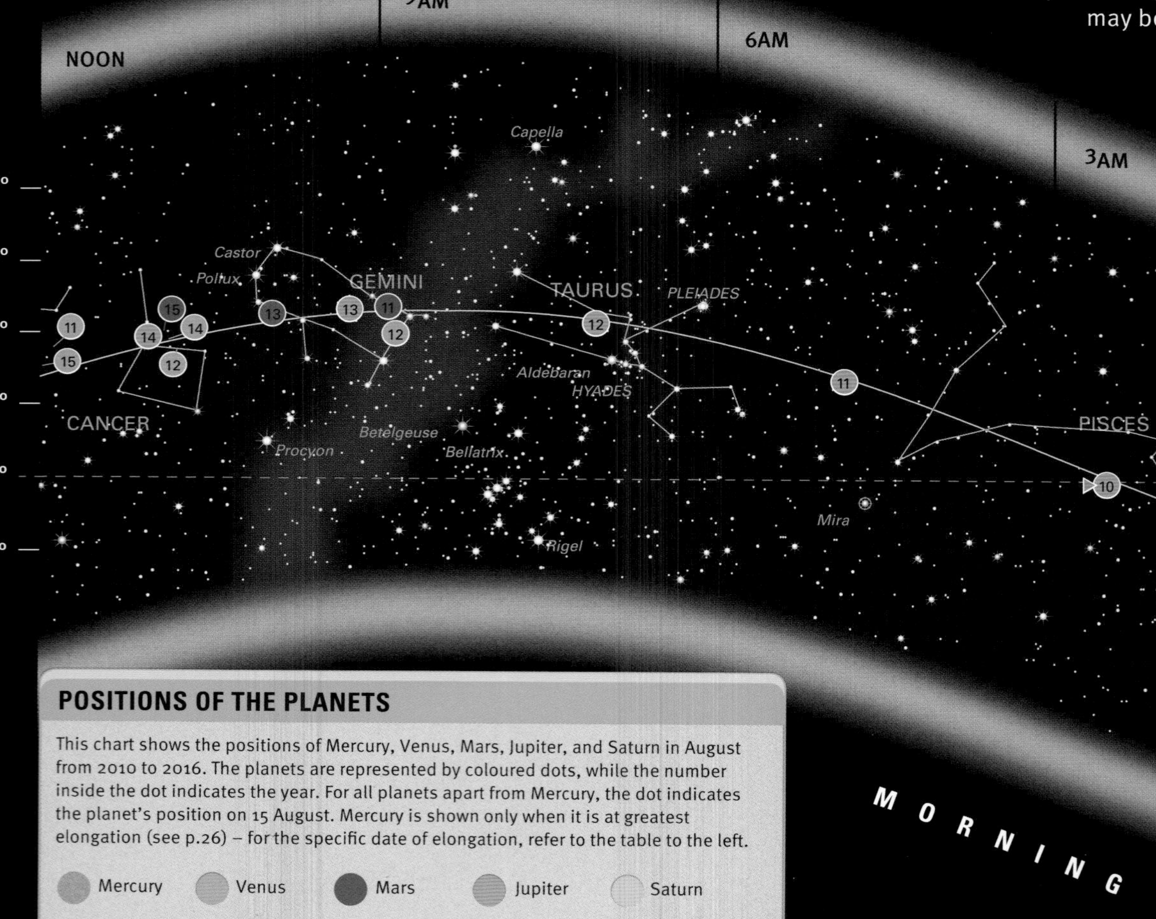

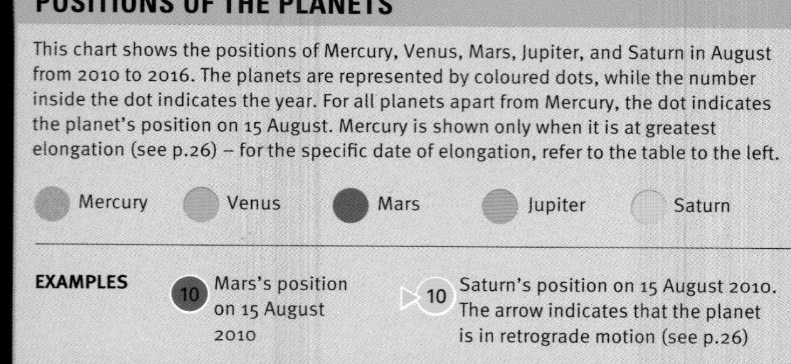

POSITIONS OF THE PLANETS

This chart shows the positions of Mercury, Venus, Mars, Jupiter, and Saturn in August from 2010 to 2016. The planets are represented by coloured dots, while the number inside the dot indicates the year. For all planets apart from Mercury, the dot indicates the planet's position on 15 August. Mercury is shown only when it is at greatest elongation (see p.26) – for the specific date of elongation, refer to the table to the left.

Mercury | Venus | Mars | Jupiter | Saturn

EXAMPLES
Mars's position on 15 August 2010
Saturn's position on 15 August 2010. The arrow indicates that the planet is in retrograde motion (see p.26)

SOUTHERN LATITUDES

Three bright stars dominate the scene to the north, the highest of which is Altair in Aquila. Below it are Vega in Lyra, and Deneb in Cygnus. The three form a triangle known as the Summer Triangle due to its prominence in the northern-hemisphere summer sky. The constellation of Vulpecula, the fox, which is between Aquila and Cygnus, contains a relatively easy-to-spot planetary nebula, the Dumbbell Nebula (M27). It is seen through binoculars as a rounded patch of light.

A second well-known planetary nebula, the Ring Nebula (M57), is in Lyra and can be seen through a small telescope as a disc of light. The constellation Tucana, the toucan, is to the southeast. It contains the globular cluster 47 Tucanae and the irregular galaxy called the Small Magellanic Cloud.

Sagittarius and the rich star fields in the centre of the galaxy are still high overhead. The galaxy's milky path flows through Scorpius to the southwest, then through Lupus and on to Crux and the southern horizon. Alpha (α) and Beta (β) Centauri are low on the southwestern horizon and mark the front legs of Centaurus, the centaur (half man, half horse). Alpha Centauri, also known as Rigil Kentaurus, is the third-brightest star in the sky; Beta Centauri, or Hadar, is 11th brightest. The two appear to be the same distance from us, but in reality they are vastly separated. Beta is about 525 light years away, and Alpha only 4.3 light years away. A small telescope will show that Alpha Centauri is a binary star consisting of two yellow and orange stars that orbit each other every 80 years. Alpha Centauri is sometimes described as the closest star after the Sun, but Proxima Centauri, a much fainter red dwarf star possibly associated with Alpha, is a little closer. Centaurus is one of two centaurs in the sky; the second is Sagittarius.

47 Tucanae
The globular cluster 47 Tucanae is the second brightest in the sky. At 13,400 light years away, it is also one of the closest to Earth. It contains several million stars and appears as a hazy star to the naked eye.

Cygnus Rift
The path of the Milky Way flows from Cepheus (top) to Scorpius (near the horizon), where it is noticeably brighter as we look towards the galaxy's centre. The dark band of dust along its centre line is the Cygnus Rift.

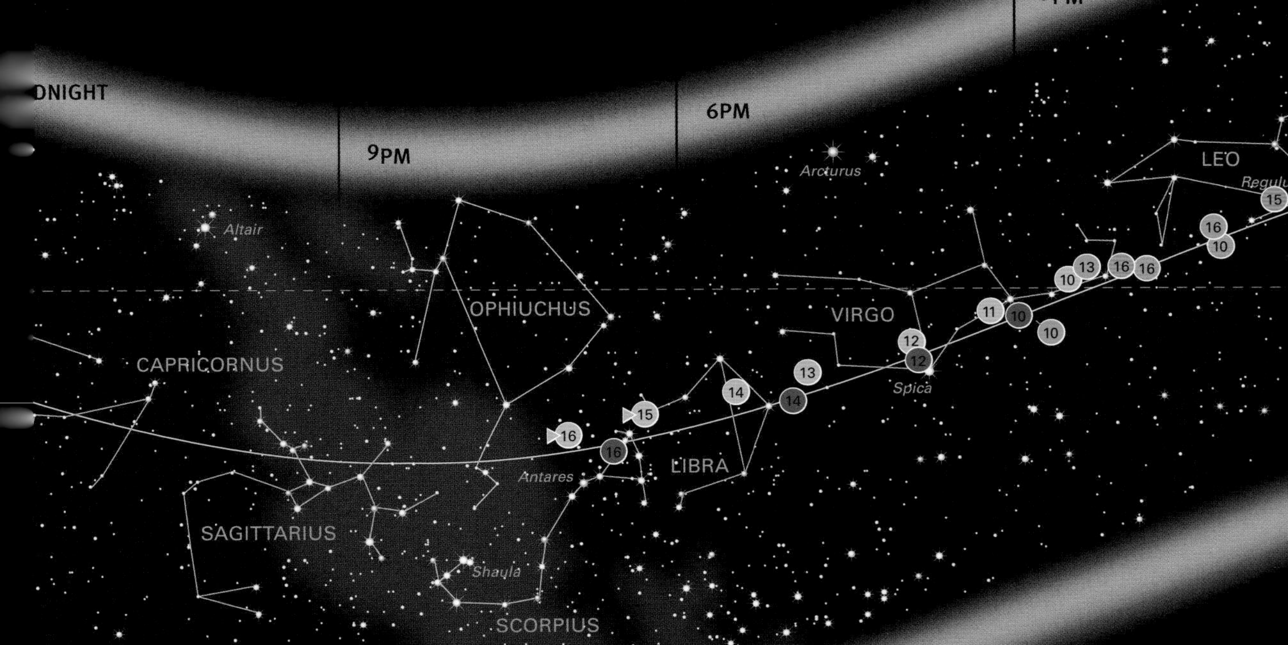

September

The stars of northern autumn and southern spring are now in place as Capricornus and Aquarius move centre stage. On the 23rd, or occasionally the 22nd, of the month, the Sun moves from the northern to the southern celestial sky. When it crosses the celestial equator, day and night are of equal length. In the weeks ahead, the northern hemisphere nights lengthen and the southern ones shorten.

SPECIAL EVENTS

PHASES OF THE MOON

	Full Moon	New Moon
2010	23 September	8 September
2011	12 September	27 September
2012	30 September	16 September
2013	19 September	5 September
2014	9 September	24 September
2015	28 September	13 September
2016	16 September	1 September

ECLIPSES

2015: 13 September A partial solar eclipse is visible from southern Africa and parts of Antarctica.

2015: 28 September A total eclipse of the Moon is visible from the Americas, Europe, and Africa.

2016: 1 September An annular eclipse of the Sun is visible from the Atlantic Ocean, central Africa, Madagascar, and the Indian Ocean.

THE PLANETS

2010: 19 September Mercury is at greatest morning elongation, magnitude –0.1.

2010: 21 September Jupiter is at opposition, magnitude –2.9. At midnight, it is visible to the south from northern latitudes and to the north from southern latitudes.

2011: 3 September Mercury is at greatest morning elongation, magnitude 0.0.

2013: 20 September Venus and Saturn appear seven Moon widths apart in the western evening sky.

2014: 21 September Mercury is at greatest evening elongation, magnitude 0.1.

2015: 4 September Mercury is at greatest evening elongation, magnitude 0.3.

2016: 28 September Mercury is at greatest morning elongation, magnitude –0.4.

Opposition and **elongation** are explained on *page 26*

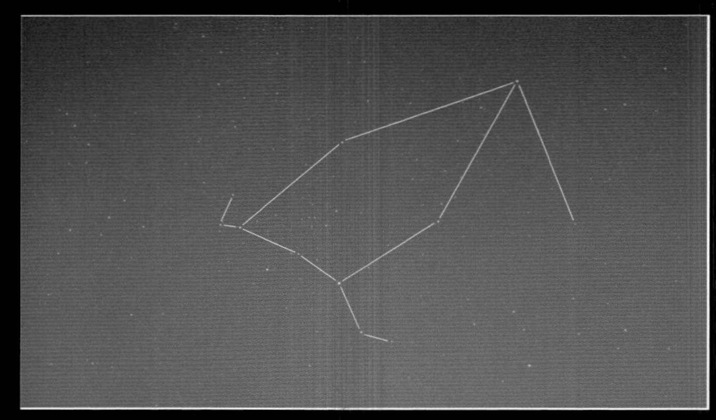

Cepheus
The head of Cepheus, the mythical king of Ethiopia, is at lower left. The constellation is not prominent but is worth seeking out for Delta Cephei, in the king's head. Its brightness changes between magnitudes 3.5 and 4.4.

NORTHERN LATITUDES

Capricornus and Aquarius are at either side of due south. For those at high northern latitudes, September is the best time to see Capricornus, the smallest zodiac constellation. Cygnus remains high overhead and is joined by the flying horse, Pegasus. The Summer Triangle stars, Deneb (Cygnus), Vega (Lyra), and Altair (Aquila), remain in view to the west, but Pegasus's presence in the east announces the arrival of autumn.

To the north, Ursa Major is below Polaris, while Cepheus is above. Its star Delta (δ) Cephei is the prototype of the Cepheid variables. As the star pulsates, its brightness shifts over a five-day cycle. Only two bright stars are seen to the north: Capella (Auriga), the sixth brightest of all, is close to the northeast horizon. Just brighter is Vega, high in the sky to the west.

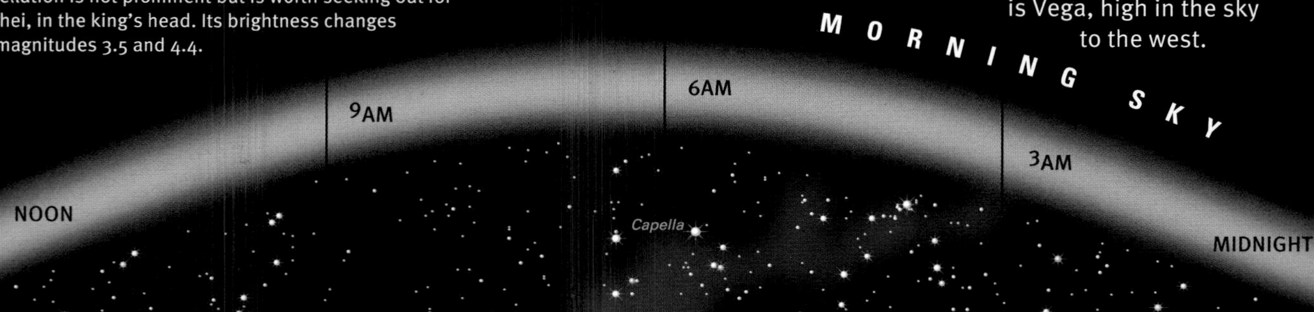

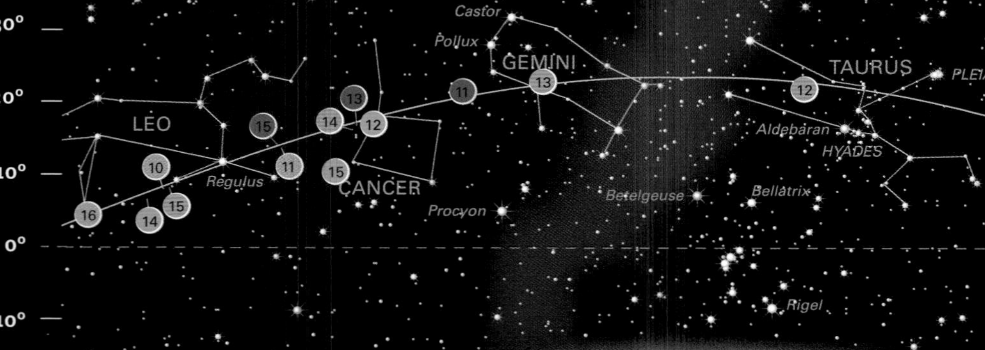

POSITIONS OF THE PLANETS

This chart shows the positions of Mercury, Venus, Mars, Jupiter, and Saturn in September from 2010 to 2016. The planets are represented by coloured dots, while the number inside the dot indicates the year. For all planets apart from Mercury, the dot indicates the planet's position on 15 September. Mercury is shown only when it is at greatest elongation (see p.26) – for the specific date of elongation, refer to the table to the left.

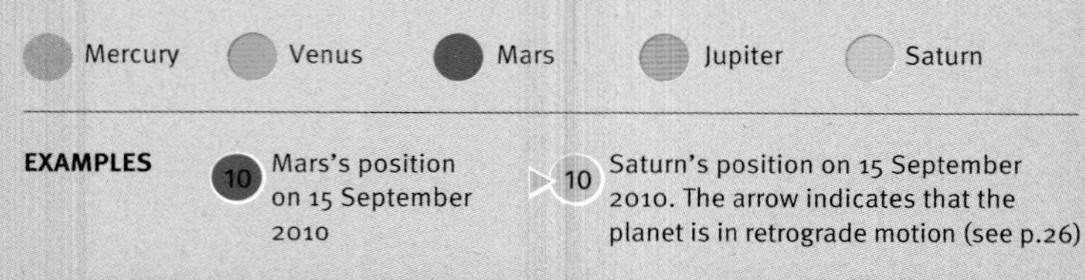

Small Magellanic Cloud
The Small Magellanic Cloud (left) and the globular cluster 47 Tucanae (right) seem the same distance from us. In reality, the galaxy, which is 10,000 light years across, is 210,000 light years away. 47 Tucanae is just 13,400 light years away.

SOUTHERN LATITUDES

Capricornus and Aquarius are overhead. Neither has especially bright stars, but Aquarius in particular has objects of note. The brightest star in Capricornus, Alpha (α), called Algedi, is a double star. Sharp eyesight or binoculars show a giant of magnitude 3.6 and, six times more distant, a supergiant of magnitude 4.3.

Aquarius, the water carrier, includes M2, a globular cluster that is easily seen as a fuzzy star through binoculars, and a notable planetary nebula. The Helix Nebula (NGC 7293) is believed to be the closest planetary nebula to Earth. It appears large in the sky and covers a third of the apparent width of the Moon. As a result its light is spread out, so it is difficult to spot. Binoculars will locate its pale-grey light. Close by is the bright star Fomalhaut in Piscis Austrinus.

Altair (Aquila), Vega (Lyra), and Deneb (Cygnus) form the Summer Triangle and are still visible to the northwest. September offers the last chance of seeing the triangle before Vega and, later, Deneb move below the horizon. Take a last look at Albireo, in the beak of the swan. It is a celebrated double that is clearly separated through a telescope.

The path of the Milky Way is seen in the western sky. Sagittarius remains high; below is Scorpius. Lowest of all are Centaurus and Crux, which are disappearing below the southwestern horizon. The blue-white star Achernar is well placed to the southeast. At magnitude 0.45, it is the ninth brightest of all. Achernar marks one end of the mythical river Eridanus. Its name, Arabic in origin, means “river’s end”.

To the right of Achernar is Tucana, representing the tropical bird, the toucan. It is home to the globular cluster 47 Tucanae (NGC 104). Close by is an elongated patch of light seven times wider than the apparent width of the full Moon. This is the Small Magellanic Cloud – a small irregular galaxy that is a companion to the Milky Way. When viewed with binoculars, this naked-eye object reveals some of its star clusters and nebulae.

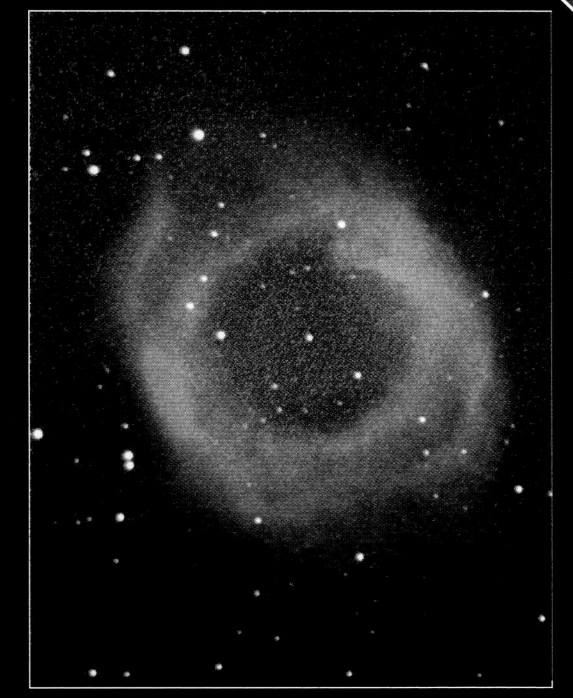

Helix Nebula
The colour and complexity of the Helix Nebula in Aquarius is seen only in images such as this one. The shells of gas, which are about 15 light years across, are material expelled by a central dying star. The nebula is about 300 light years away.

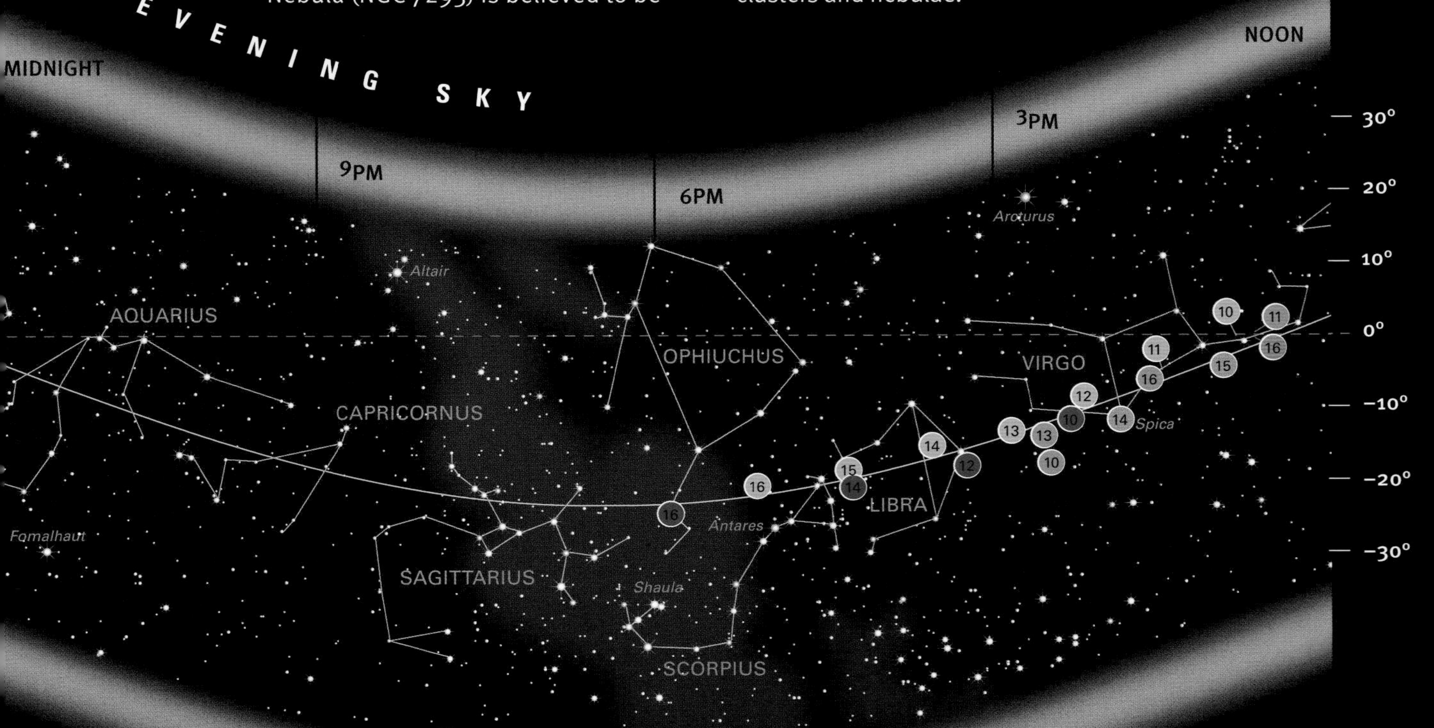

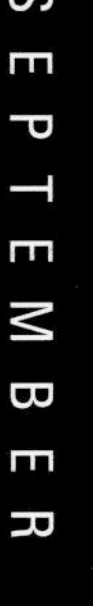

October

The Andromeda Galaxy, the most distant object visible to the naked eye, is on view to all observers as is the Great Square of Pegasus, an asterism formed by linking stars in neighbouring Pegasus an Andromeda. Southern observers can see both of the Milky Way's companion galaxies in the sky, and Cassiopeia is well positioned for observers in northern latitudes.

SPECIAL EVENTS

PHASES OF THE MOON

	Full Moon	New Moon
2010	23 October	7 October
2011	12 October	26 October
2012	29 October	15 October
2013	18 October	5 October
2014	8 October	23 October
2015	27 October	13 October
2016	16 October	1, 30 October

ECLIPSES

2014: 8 October A total eclipse of the Moon is visible from east Asia and North America.

2014: 23 October A partial solar eclipse is visible from central and western USA, Canada, and Mexico.

THE PLANETS

2011: 29 October
Jupiter is at opposition, magnitude –2.9. At midnight, it is visible to the south from northern latitudes and to the north from southern latitudes.

2011: 30–31 October
Mercury and Venus appear four Moon widths apart in the western evening sky.

2012: 4–5 October
Mercury and Saturn appear about six Moon widths apart in the western evening sky.

2012: 26 October
Mercury is at greatest evening elongation, magnitude –0.1.

2013: 9 October
Mercury is at greatest evening elongation, magnitude 0.0.

2013: 10 October
Mercury and Saturn appear 11 Moon widths apart in the southwestern evening sky.

2015: 16 October,
Mercury is at greatest morning elongation, magnitude –0.5.

2015: 17–30 October
Venus, Jupiter, and Mars appear within two Moon widths in the eastern morning sky.

2016: 30 October
Venus and Saturn appear six Moon widths apart in the southwestern evening sky.

Opposition and **elongation** are explained on *page 26*

NORTHERN LATITUDES

Pegasus and Andromeda are either side of due south. The winged horse and the princess are linked by the Great Square of Pegasus. This is formed of three stars in Pegasus and one in Andromeda, and it defines the upper body of the mythological winged horse. Close to the horse's nose is M15, a globular cluster visible to the naked eye under clear rural skies.

Andromeda is home to the Andromeda Galaxy (M31), the largest member of the Local Group cluster. It is a spiral but is tipped to our line of sight, so it looks like an elongated oval. Its central part can be seen with the naked eye and binoculars. A large telescope is needed for its spiral arms.

The Orionid meteor shower peaks around 20 October, with about 25 fast meteors an hour. It is best seen after midnight when Orion has risen in the east.

Looking northwards, Vega (in Lyra), Polaris (which marks the position of the north celestial pole), and Capella (in Auriga) make across the sky. The path of the Milky Way arches overh Cepheus and Cassiopeia are above Polaris and are in t optimum positions for viewing.

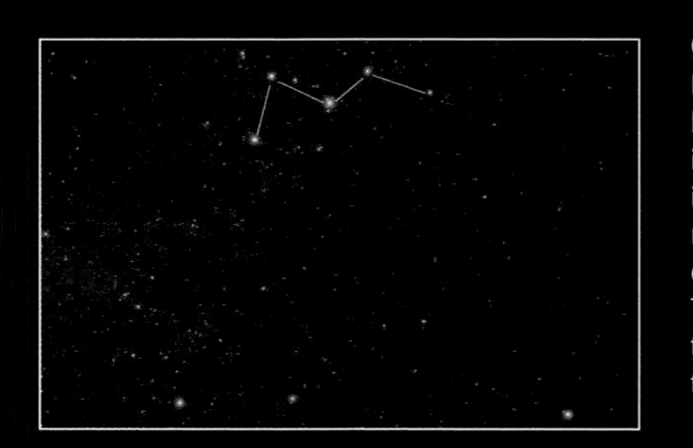

Cassiopeia
Here, the "M" shape bright stars in Cassio is at upper centre. P is the bright star at b right, and two bright of Cepheus are to its The Milky Way flows through Cassiopeia top right to bottom l

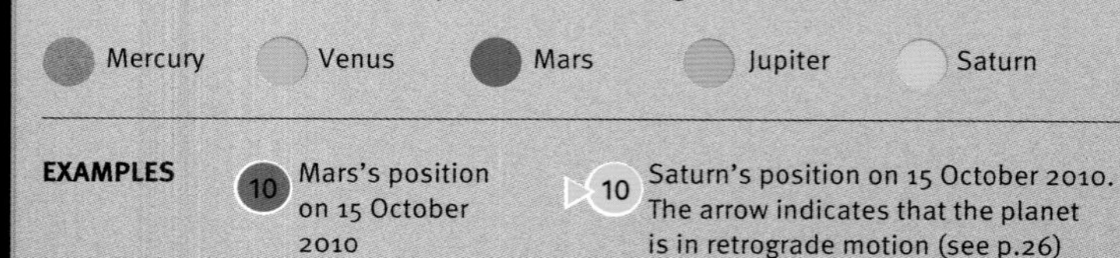

POSITIONS OF THE PLANETS

This chart shows the positions of Mercury, Venus, Mars, Jupiter, and Saturn in October from 2010 to 2016. The planets are represented by coloured dots, while the number inside the dot indicates the year. For all planets apart from Mercury, the dot indicates the planet's position on 15 October. Mercury is shown only when it is at greatest elongation (see p.26) – for the specific date of elongation, refer to the table to the left.

Mercury   Venus   Mars   Jupiter   Saturn

EXAMPLES Mars's position on 15 October 2010. Saturn's position on 15 October 2010. The arrow indicates that the planet is in retrograde motion (see p.26)

SOUTHERN LATITUDES

Looking north, the Great Square of Pegasus is centrally placed. It is formed by three stars in the winged horse Pegasus, to the left, and one star in the mythological princess, Andromeda, to the right. The horse flies towards the west. His head is defined by a line of stars; below, two more lines map out his forelegs. The square makes the shape of his upper body.

Andromeda is lower in the sky and to the right of Pegasus. Alpheratz, its brightest star, marks the princess's head and is the fourth star in the Great Square. The Andromeda Galaxy (M31) is in her left knee. This spiral – similar to but larger than the Milky Way and 2.5 million light years away – appears as an elongated oval to the naked eye. Its spiral arms and two companion galaxies, M32 and M110, are seen through a larger telescope. A small telescope will reveal NGC 7662, the Blue Snowball planetary nebula near Andromeda's right hand.

The bright star Fomalhaut, in Piscis Austrinus, is almost overhead. The winter trio of stars Altair, Vega, and Deneb is setting in the northwest. The summer constellations Taurus and Orion are beginning to appear in the east. Only one bright star, Achernar (Eridanus), is high above the horizon to the south. Four relatively dim constellations depicting birds – Phoenix, Grus, Tucana, and Pavo – are centre stage. Tucana is well placed for observing the Small Magellanic Cloud and 47 Tucanae. Close to the southeastern horizon is Dorado, significant because it contains the Large Magellanic Cloud, the larger and closer of our two companion galaxies.

Large Magellanic Cloud
Binoculars and a small telescope reveal star clusters and nebulous gas and dust patches in the Large Magellanic Cloud. On its upper-left edge is the Tarantula Nebula, including a cluster of new-born stars. The Large Magellanic Cloud appears to the naked eye as a long hazy patch of light.

Pegasus, the winged horse
The Great Square of Pegasus forms the upper body of the horse. His head is here below the square, and his forelegs stretch off the image at the right. The star at left in the square belongs to Andromeda. The Moon is seen close to the horizon, and above it is the planet Venus.

November

Two celebrated stars – Mira in Cetus, the sea monster, and Algol in Perseus – are in the sky for both northern and southern observers. Cetus and Perseus are joined by the constellation of Andromeda, the princess rescued from Cetus by Perseus. Northern-latitude sky watchers see the Milky Way arching overhead, and southern-latitude observers see four birds as they await the arrival of the summer stars.

SPECIAL EVENTS

PHASES OF THE MOON

	Full Moon	New Moon
2010	21 November	6 November
2011	10 November	25 November
2012	28 November	13 November
2013	17 November	3 November
2014	6 November	22 November
2015	25 November	11 November
2016	14 November	29 November

ECLIPSES

2011: 25 November A partial eclipse of the Sun is visible from the southern Indian Ocean and Antarctica.

2012: 13–14 November A total eclipse of the Sun is visible from northeastern Australia and the South Pacific. A partial eclipse is visible from New Zealand, the rest of Australia, and the Pacific Ocean.

2013: 3 November A total eclipse of the Sun is visible from the mid Atlantic Ocean and central Africa.

THE PLANETS

2010: 20 November Mercury and Mars appear about three Moon widths apart in the western evening sky.

2011: 1–14 November Mercury and Venus appear four Moon widths apart in the western evening sky.

2011: 14 November Mercury is at greatest evening elongation, magnitude –0.2.

2012: 27 November Venus and Saturn appear one Moon width apart in the eastern dawn sky.

2013: 18 November Mercury is at greatest morning elongation, magnitude –0.5.

2013: 26 November Mercury and Saturn appear one Moon width apart in the eastern morning sky.

2014: 1 November Mercury is at greatest morning elongation, magnitude –0.5.

2016: 24 November Mercury and Saturn appear seven Moon widths apart in the southwestern evening sky.

Elongation is explained on *page 26*

NORTHERN LATITUDES

Perseus and Andromeda are high overhead. Andromeda's parents, Cepheus and Cassiopeia, are to the north. Andromeda is well placed for observing the Andromeda Galaxy. Algol in Perseus is an eclipsing binary – two stars that exist together and orbit each other. As the dimmer of the pair passes in front of the brighter, the combined brightness drops. Algol's magnitude changes from 2.1 to 3.4 and back again.

Cetus is due south. Its red giant Mira is a variable star whose brightness varies in a cycle lasting 11 months. It changes from a naked-eye star of magnitude 3, to a star of magnitude 10 visible only through a telescope. Pisces is above Cetus; to its right is the Great Square of Pegasus.

The Summer Triangle stars of Altair (Aquila), Vega (Lyra), and Deneb (Cygnus) are low in the northwest. The winter constellations Taurus, Gemini, and Orion are to the southeast. There are two meteor showers this month: the Taurids peak in the first week and the Leonids around the 17th. About ten meteors an hour are seen in each.

Leonid meteor shower
Thirty images are combined in this view of a Leonid meteor shower. Meteors are best seen about 50 degrees above the horizon and 30–40 degrees to one side of the point from which they appear to radiate – in this case, in Leo.

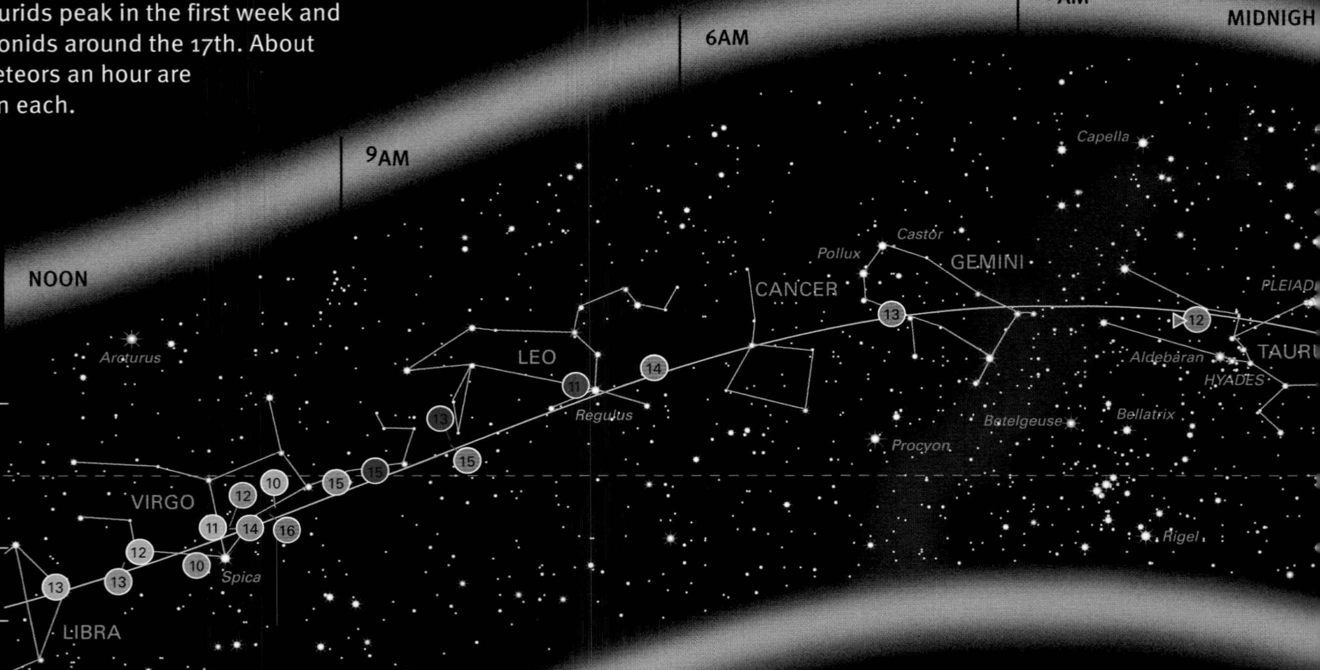

POSITIONS OF THE PLANETS

This chart shows the positions of Mercury, Venus, Mars, Jupiter, and Saturn in Novemb from 2010 to 2016. The planets are represented by coloured dots, while the number inside the dot indicates the year. For all planets apart from Mercury, the dot indicates the planet's position on 15 November. Mercury is shown only when it is at greatest elongation (see p.26) – for the specific date of elongation, refer to the table to the left.

Mercury Venus Mars Jupiter Saturn

EXAMPLES

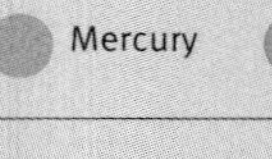
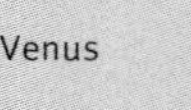
Mars's position on 15 November 2010

Saturn's position on 15 November 20 The arrow indicates that the planet is in retrograde motion (see p.26)

Mira
The variable star Mira is circled in the centre of this image. It is here at its brightest and visible to the naked eye. The two brightest "stars" in this view are, in fact, planets: Jupiter is at right, with Saturn above and to its left.

SOUTHERN LATITUDES

The constellation Cetus is overhead. Mira, one of its stars, is a well-known variable. It is a red giant whose brightness changes over about 11 months. It is about magnitude 3 at brightest and within naked-eye visibility. At its dimmest, about magnitude 10, it is visible only through a telescope. Mira is the prototype for long-period variables known as Mira variables.

Pisces and Aries are either side of north. Pisces, depicting two fish joined with a cord, is a faint constellation. Its brightest star, alpha (α), marks the knot that ties the fish together; a telescope reveals it as two stars. Below Pisces and to the northeast are Pegasus and Andromeda, which are linked by the Great Square of Pegasus. The Andromeda Galaxy remains high enough to be seen. November offers the chance to see Andromeda and Perseus together. Perseus, who killed the sea monster Cetus and prevented it from devouring Andromeda, is near the horizon in the northeast. The star Beta (β) Persei, called Algol, is a well-known eclipsing binary. It consists of two stars orbiting around each other. Together they are magnitude 2.1, but when the fainter passes in front of its brighter companion, the brightness drops to magnitude 3.4. This change can be detected by the naked eye. The dip and return to full brightness takes about 10 hours every 69 hours.

Achernar in Eridanus is centre stage to the south. To its west is Fomalhaut in Piscis Austrinus, the southern fish. The fish is the parent of the two in Pisces and is the recipient of water pouring from Aquarius's jug at right. Four birds – Phoenix, Grus (the crane), Tucana (the toucan), and Pavo (the peacock) – are in the southwest. The Small Magellanic Cloud is seen in Tucana, and to its left is the Large Magellanic Cloud in Dorado. Both irregular galaxies are visible to the naked eye. Individual clusters and nebulae are seen with binoculars or a telescope. Canopus, the white supergiant and second-brightest star of all, is in the southeast. Beyond, in the east, is Sirius (Canis Major), the brightest of all. The presence of Taurus, Orion, and Canis Major above the horizon is a sign that summer is approaching.

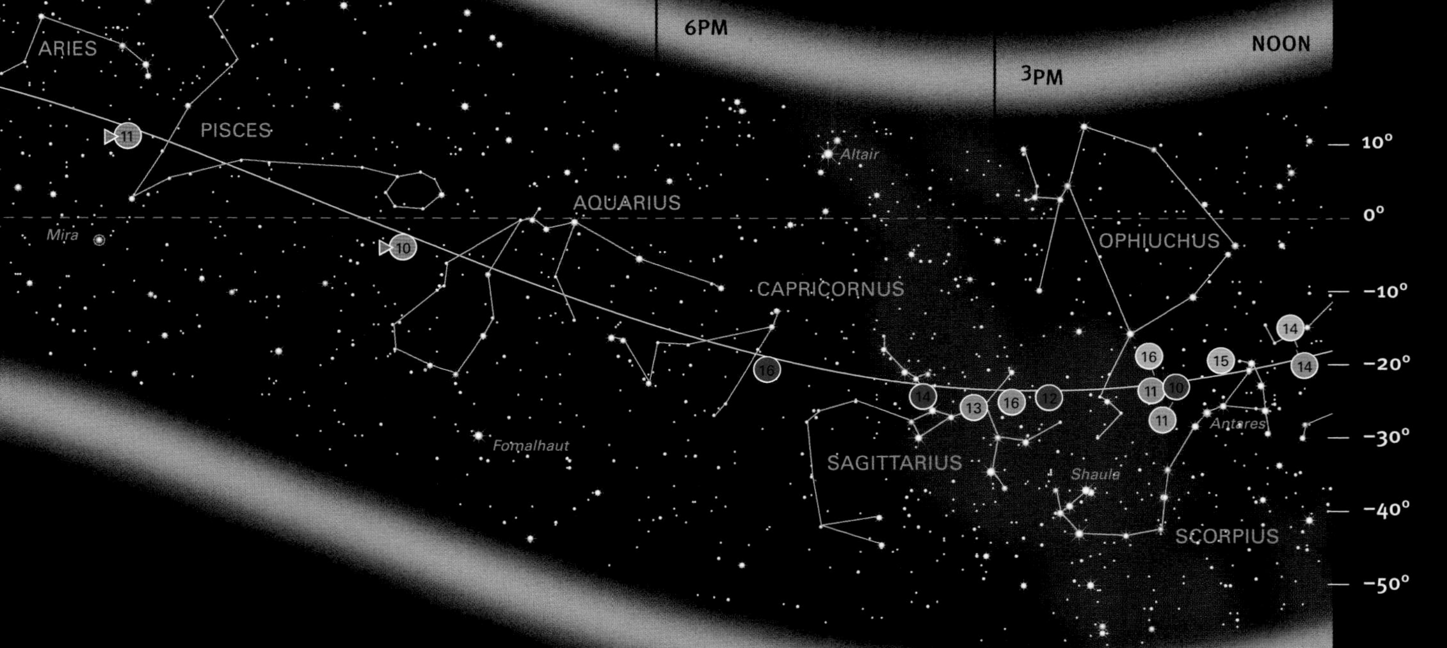

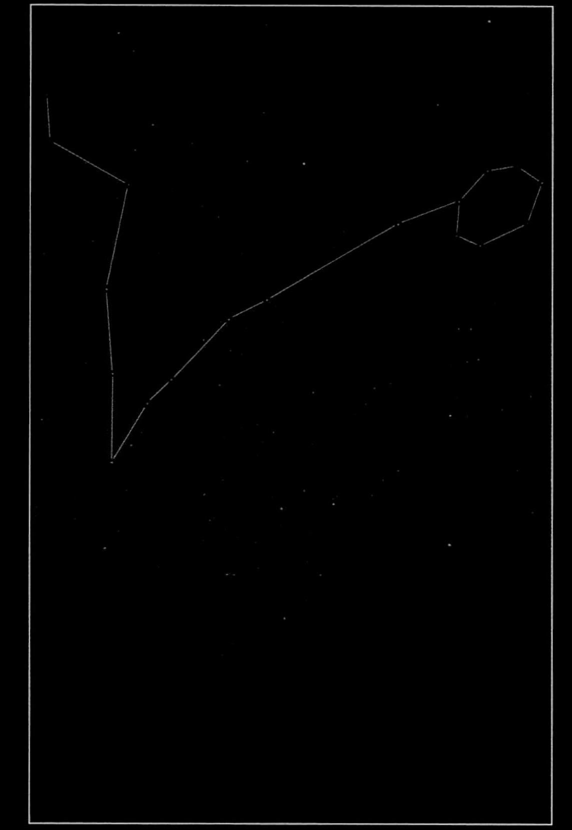

Pisces
The distinguishing feature of Pisces is the ring of seven stars that marks the body of one of the two fish. Alpha (α) Piscium (lower left) marks the knot where the cords from the fish are tied together.

December

Taurus is ideally placed for everyone to observe its fine star clusters. This month also offers the first opportunity for a good look at Orion as it moves higher in the sky. The Sun is at its farthest point south of the celestial equator on 21–22 December. Northern-latitude nights are the longest of the year, while southern ones are the shortest.

SPECIAL EVENTS

PHASES OF THE MOON

	Full Moon	New Moon
2010	21 December	5 December
2011	10 December	24 December
2012	28 December	13 December
2013	17 December	3 December
2014	6 December	22 December
2015	25 December	11 December
2016	14 December	29 December

ECLIPSES

2010: 21 December A total eclipse of the Moon is visible from North and South America and the Pacific Ocean.

2011: 10 December A total eclipse of the Moon is visible from the Pacific Ocean, Australasia, and Asia.

THE PLANETS

2010: 1 December
Mercury is at greatest evening elongation, magnitude –0.4.

2011: 23 December
Mercury is at greatest morning elongation, magnitude –0.3.

2012: 3 December
Jupiter is at opposition, magnitude –2.8. At midnight, it is visible to the south from northern latitudes and to the north from southern latitudes.

2012: 4 December
Mercury is at greatest morning elongation, magnitude –0.3.

2014: 30 December
Mercury and Venus appear seven Moon widths apart in the southwestern evening sky.

2015: 29 December
Mercury is at greatest evening elongation, magnitude –0.5.

2016: 11 December
Mercury is at greatest evening elongation, magnitude –0.4.

Opposition and **elongation** are explained on *page 26*

NORTHERN LATITUDES

Winter stars occupy half the view to the south: Taurus leads the way and is high and almost due south; Orion is in the southeast; and Gemini in the eastern sky. Two water constellations, Pisces, the fish, and Cetus, the sea monster, move towards the western horizon. Two bright stars shine out above and below the figure of Orion: yellow Capella, in Auriga and sixth brightest, shines overhead; brilliant-white Sirius, the brightest of all, is near the southeast horizon.

Taurus is easy to see due to its distinctive shape. Its stars depict a bull's head and shoulders; the bright red giant Aldebaran is its eye. Aldebaran appears to be part of the Hyades open cluster in the bull's face but is only half its distance away. The Hyades are easily visible to the naked eye; binoculars reveal many more stars. A second cluster, the Pleiades (M45), marks the start of the bull's back. About six stars are seen by eye; dozens through binoculars. A telescope shows the Crab Nebula (M1) next to the tip of a horn. It is the remains of a star that exploded in 1054.

To the north, both Ursa Major and Ursa Minor are below Polaris; Andromeda and Perseus remain overhead. A naked-eye knot of light in Perseus's hand is the Double Cluster. The Geminids, the second-best meteor shower, reaches its peak on the 13th, when up to 100 meteors an hour are seen.

Double Cluster
The Double Cluster is a pair of open star clusters within Perseus. NGC 884 is to the left, and NGC 869 to the right. Each contains hundreds of stars. The naked eye sees them as a brighter patch in the path of the Milky Way.

Canopus

The second-brightest star in the entire sky, Canopus is a white supergiant, 310 light years away. At magnitude –0.62, it is easily visible to the naked eye. It is a truly brilliant star – placed next to the Sun, it would outshine it 14,000 times.

SOUTHERN LATITUDES

Looking north, the spring constellations Aquarius, Pisces, and Pegasus are moving towards the western horizon as the summer stars rise in the east. The upside-down figure of Orion, the hunter, announces that summer is almost here. It is flanked by Taurus to its lower left and Canis Major to the right. Below is the constellation Gemini, the twins. The two bright stars that mark their heads, Castor and Pollux, are near the northeastern horizon. Their legs point upwards to Orion.

Taurus is the front portion of a bull, which, from southern latitudes, is seen facing east, the direction it has come from. Its shoulders and head are close to the horizon, and its forelegs point up to the sky overhead. The Pleiades (M45), one of the finest open clusters in the sky, marks the bull's shoulder. Six of its stars are discernible with the naked eye; the keen-sighted may see seven, which gives rise to the cluster's alternative name, the Seven Sisters. Many more are visible with optical aid. The bull's face is formed of a second, V-shaped cluster, the Hyades. More than a dozen of its stars are visible to the naked eye. The unrelated red star Aldebaran represents one eye.

Eridanus, the river, is overhead. Its meandering path is traced from Orion's bright star, Rigel, a blue supergiant of magnitude 0.18. It flows overhead, then towards the southern horizon, stopping at Achernar at right of due south. Brilliant-white Canopus (in Carina) is to the left of Achernar. Between them, but closer to the horizon, are two galaxies: the Large and Small Magellanic Clouds.

The brightest star of all, Sirius in Canis Major, shines high in the east. Sirius is one corner of a triangle of stars in the east; the other two are Betelgeuse in Orion and Procyon in Canis Minor.

Hyades and Pleiades

The V-shaped group of stars below and left of centre is the Hyades star cluster representing the face of Taurus. The brightest star in the face is Aldebaran, which is not part of the cluster. The knot of stars at centre right is the Pleiades cluster.

POSITIONS OF THE PLANETS

This chart shows the positions of Mercury, Venus, Mars, Jupiter, and Saturn in December from 2010 to 2016. The planets are represented by coloured dots, while the number inside the dot indicates the year. For all planets apart from Mercury, the dot indicates the planet's position on 15 December. Mercury is shown only when it is at greatest elongation (see p.26) – for the specific date of elongation, refer to the table on the facing page.

Mercury Venus Mars Jupiter Saturn

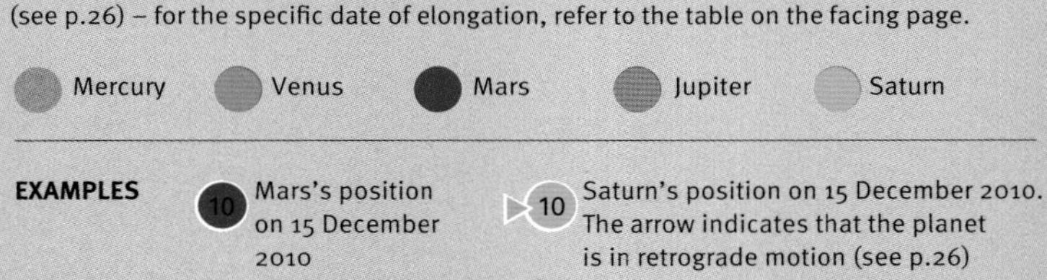

EXAMPLES Mars's position on 15 December 2010

Saturn's position on 15 December 2010. The arrow indicates that the planet is in retrograde motion (see p.26)

Glossary

Words in *italics* are defined elsewhere in the glossary.

ABSOLUTE MAGNITUDE A measure of the actual brightness of an object, defined as the *apparent magnitude* it would have at a distance of 32.6 *light years*.

APERTURE The diameter of the main mirror or lens in a telescope or binoculars. A large aperture telescope can see more detail and detect fainter objects than a small aperture telescope.

APPARENT MAGNITUDE The brightness of a celestial object as seen from the Earth. This depends on the object's real brightness and its distance from the Earth.

ASTERISM A pattern of *stars* where the *stars* are either a part of a *constellation*, or are members of several *constellations*. An example is the Plough in Ursa Major.

ASTEROID A rocky body orbiting the Sun with a diameter less than 1,000 km.

ASTRONOMICAL UNIT (AU) A measure of distance convenient for use within the Solar System, defined as the mean distance between the Earth and the Sun (149,597,970 km).

ASTROPHOTOGRAPHY Photography of celestial objects in the night sky; also includes photography of the Sun and *eclipses*.

BINARY STAR Two *stars* in mutual *orbit* around a common centre of mass, and bound together gravitationally.

CELESTIAL EQUATOR The celestial equivalent of the Earth's equator. The celestial equator marks a line where the plane of the Earth's equator meets the *celestial sphere*.

CELESTIAL POLE The celestial equivalent of the Earth's poles. The night sky appears to rotate on an axis through the celestial poles.

CELESTIAL SPHERE The imaginary sphere that surrounds the Earth, and upon which all celestial objects appear to lie.

CEPHEID VARIABLE A type of *variable star* with a regular pattern of brightness changes linked to the *star*'s actual *luminosity*. Such *stars* are used as distance indicators.

CHARGE-COUPLED DEVICE (CCD) A light-sensitive silicon chip used as an alternative to photographic film.

COMET An icy body orbiting the Sun that may develop a glowing tail as it passes through the inner Solar System.

CONJUNCTION An alignment of objects in the night sky, with one passing in front of the other, particularly when a planet lines up with the Sun as viewed from the Earth.

CONSTELLATION An area of the night sky with boundaries that are determined by the International Astronomical Union; constellations are 88 in number.

DECLINATION The celestial equivalent of latitude on the Earth. It is the angle between a celestial object and the *celestial equator*, measured in degrees. The *celestial equator* has a declination of 0 degrees, and the *celestial poles* are at 90 degrees.

DEEP-SKY OBJECT Any celestial object external to the Solar System, but excluding *stars*.

DIFFUSE NEBULA A cloud of gas and dust illuminated by *stars* embedded within it.

DOUBLE STAR Two *stars* that are not physically associated with each other but appear close together through line-of-sight from the Earth.

DWARF PLANET A celestial body that is large enough to be nearly round, but that has not cleared its neighbourhood of other objects.

DWARF STAR A *star* that has lost most of its mass towards the end of its evolutionary development.

ECCENTRICITY A measure of the circularity of a body's *orbit*. An eccentricity of 0 means a circular orbit, with larger values indicating more elongated ellipses up to a theoretical maximum of 1.

ECLIPSE An alignment of a planet or moon with the Sun, which casts a shadow on another body. During a lunar eclipse, the Earth's shadow is cast on the Moon. During a solar eclipse, the Moon's shadow is cast on the Earth.

ECLIPTIC The plane of the Earth's *orbit* around the Sun, or the projection of that plane onto the *celestial sphere*.

ELONGATION The angular separation between the Sun and a planet as viewed from the Earth. Also used as the time of maximum angular separation (greatest elongation) between the inner planets, Mercury or Venus, and the Sun.

GALAXY A huge mass of *stars*, gas, and dust, containing from millions to billions of *stars*. Galaxies vary in size and shape, with diameters that range from thousands to hundreds of thousands of *light years*.

GIANT STAR A *star* that has expanded dramatically as it nears the end of its life-cycle.

GLOBULAR CLUSTER A sphere of *stars*, bound together gravitationally, and containing from tens of thousands to hundreds of thousands of *stars*.

LIBRATION A monthly variation in the parts of the Moon's surface visible from the Earth, due to the slight *eccentricity* and tilt of the Moon's *orbit*.

LIGHT YEAR The distance light can travel during the course of one year, that is, 9,460,700,000,000 km (5,878,600,000,000 miles).

LIMB The outer edge of a moon's or a planet's observed disc.

LOCAL GROUP A small cluster of over 30 galaxies which includes our own *galaxy*, the *Milky Way*.

LONG-EXPOSURE PHOTOGRAPHY Photography of the night sky where the camera shutter remains open, often for hours, in order to record very faint objects.

LUMINOSITY A measure of the amount of light that is produced by a celestial object.

MAGNITUDE The brightness of a celestial object, measured on a numerical scale, where brighter objects are given small or negative magnitude numbers, and fainter objects are given larger magnitude numbers.

MARE (plural: maria) Dark, low-lying areas of the Moon, flooded with lava, derived from the Latin word for "sea".

METEOR A small rock that burns due to friction as it enters the Earth's atmosphere.

METEORITE A *meteor* that reaches the Earth's (or another planet's) surface.

MILKY WAY A faint band of light visible on clear dark nights, consisting of millions of *stars*; the common name for our *galaxy*.

MULTIPLE STAR A system of *stars* that are bound together gravitationally and are in mutual *orbits*. Multiple *stars* consist of at least three *stars* and up to about a dozen *stars*.

NEBULA A cloud of gas and dust, visible by either being

illuminated by embedded *stars* or nearby *stars*, or by obscuring starlight.

OPEN CLUSTER A group of up to a few hundred *stars* bound by gravity; found in the arms of a *galaxy*.

OPPOSITION The time when an outer planet lies on the exact opposite side of the Earth from the Sun. The planet is at its closest to the Earth and therefore appears brightest at this time.

ORBIT The path followed by a planet, *asteroid*, or *comet* around the Sun, or a moon around its parent planet.

PARALLAX The apparent shift in an object's position as it is viewed from two different locations. The amount of shift depends on the distance of the object, and the distance between the two locations.

PHASE Illumination of the Moon or an inner planet, as seen from the Earth. At full phase, the side of the object facing the Earth is fully illuminated; at new phase, the object is fully in shadow; crescent, half phase, and gibbous phase are in between.

PLANETARY NEBULA A shell of gas thrown off by a *star* towards the end of its evolutionary development. In a small telescope, the shell resembles a planet's disc.

PRECESSION A gradual shift in the direction of the Earth's axis of rotation. It currently points towards the *star* Polaris, but it wanders over a 25,800-year cycle.

RADIO SOURCE A celestial object that appears bright when viewed with instruments that detect radio waves.

REFLECTING TELESCOPE (reflector) A type of telescope that collects and focuses light by using a mirror.

REFRACTING TELESCOPE (refractor) A type of telescope that collects and focuses light by using a lens.

REGOLITH The loose material or "soil" on the surface of a moon or planet.

RESOLVE The ability to detect detail within celestial objects, for example, craters on the Moon, or splitting *double stars*. The greater the *aperture* of a telescope, the greater its resolving power.

RETROGRADE MOTION A reversal of the usual eastward motion of a planet relative to background *stars*; occurs as it reaches *opposition*.

RETROGRADE ROTATION The rotation of a planet or moon in the opposite direction to its *orbit*. All of the planets *orbit* the Sun in the direction of the Sun's rotation: anti-clockwise when viewed from above the Sun's north pole. Most planets also rotate (spin) anti-clockwise. Venus and Uranus have retrograde rotation: clockwise compared with their anti-clockwise *orbits*.

RIGHT ASCENSION The celestial equivalent of longitude on the Earth. It is measured in hours (one hour is 15 degrees) from the point where the Sun crosses the *celestial equator* in March.

SOLAR WIND A continuous flow of charged particles (electrons and protons) outward from the sun.

SPECTRAL TYPE A code assigned to a *star* based on the characteristics of its *spectrum*. Hot young *stars* are types O, B and A, older cooler *stars* are types F, G, K, and M.

SPECTRUM The range of wavelengths of light emitted by a celestial object, as well as any emission and absorption lines. The spectrum identifies the chemical and physical properties of the celestial object.

STAR A large sphere of gas that emits heat and light as a result of thermonuclear reactions within its core.

SUPERGIANT STAR A *star* at least ten times more massive than the Sun. Supergiants are at the end of their evolutionary development, and can be hundreds of times larger than the Sun, and thousands of times brighter.

SUPERNOVA An exceptionally violent explosion of a *star* during which it sheds its outer atmosphere; it outshines its host *galaxy*

SUPERNOVA REMNANT The outer layers of a *star* that have been ejected during a *supernova* explosion, travelling at high speed through space.

TERMINATOR The edge of the sunlit area of a moon or planet's surface, where the surface falls into shadow.

TRANSIT A planet's motion in front of the Sun, or a moon in front of its parent planet, as viewed from the Earth.

VARIABLE STAR A *star* that appears to change its brightness. This can be caused by physical changes within the *star*, or by the *star* being eclipsed by a companion.

WOLF-RAYET STAR A hot, massive *star* that produces a strong stellar wind.

ZODIAC The area of the sky, 9 degrees either side of the *ecliptic plane*, through which the Sun, the Moon, and the planets move.

Index

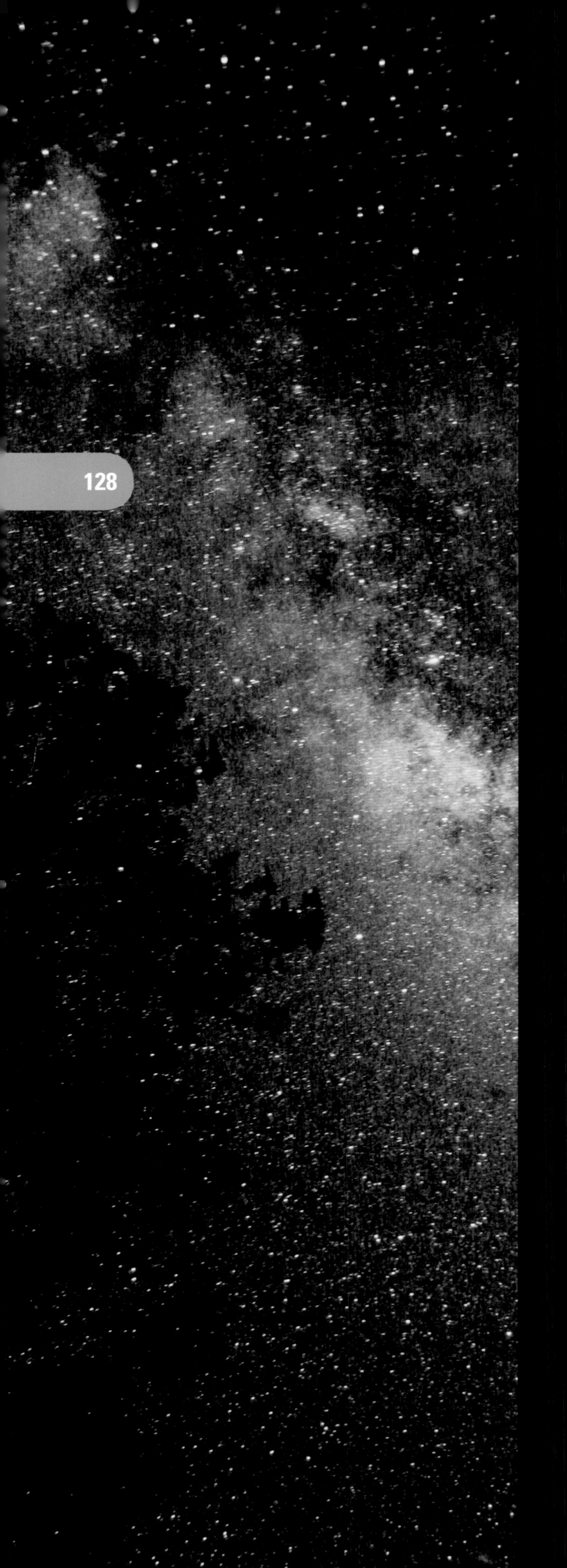

Acknowledgements

Dorling Kindersley would like to thank Heather McCarry for design consultancy and David Hughes for his contribution to the sections on Mercury, Venus, and the Moon. The planisphere artwork and the star charts on pp.32–33 were produced by Giles Sparrow and Tim Brown.

Sands Publishing Solutions would like to thank Hilary Bird for compiling the index and Robin Scagell for his assistance with images.

Photography credits

The publisher would like to thank the following for their kind permission to reproduce their photographs:

Key: a=above; b=below/bottom; c=centre; l=left; r=right; t=top
1 Science Photo Library: Fred Espenak (c).
2–3 Ali Jarekji/Reuters/Corbis: (l, r).
4–5 Corbis: Roger Ressmeyer.
6–7 HubbleSite: NASA, ESA, S. Beckwith (STScI), and the Hubble Heritage Team (STScI/AURA).
8 Corbis: Daniel J. Cox.
9 Corbis: Terra.
10–11 Martin Pugh.
12 Corbis: Tony Hallas/Science Faction.
13 Jerry Lodriguss/Astropix LLC.
14–15 Corbis: Tom Fox/Dallas Morning News.
16–17 NASA, ESA, and A Nota (STScI/ESA): (c).
18 Detlav Van Ravensway/Science Photo Library: (cr). NASA: (tl, bc, cl, c, br).
19 NASA: (bl, bc). Science Photo Library: Mark Garlick (c); Tony and Daphne Hallas (br); David A Hardy, Futures: 50 years in space (cl). Robert Williams and the Deep Field Team (STScI) and NASA: (cr).
22 Galaxy Picture Library: Robin Scagell (bl,bc,br).
24 Galaxy Picture Library: David Cortner (tr); Robin Scagell (br). NASA, ESA, and the Hubble Heritage Team (STScI/AURA) (cr).
25 Till Credner (www.allthesky.com): (bl). NOAO (br). Science Photo Library: John Chumack (bc).
26 Galaxy Picture Library: Robin Scagell (tl).
27 Galaxy Picture Library: Jon Harper (bl). Science Photo Library: Eckhard Slawik (c, tr).
28 Science Photo Library: Chris Butler (cl); Celestial Image Co (tl); Stephen and Donna O'Meara (r).
29 HST/NOAO, ESA, and the Hubble Helix Nebula Team, M Meixner (STScI), and TA Rector (NRAO): (cr). Sven Kohle (www.allthesky.com): (br). NASA, ESA, and the Hubble Heritage Team (STScI/AURA): (brr).NOAO/AURA/NSF: (brB). NOAO: (c, crr). MPIA-HD/Birkle/Slawik: (b). Science Photo Library: Celestial Image Co (car); John Chumack (tc); Eckhard Slawik. Loke Tan (www.starryscapes.com): (ca).
30 Dorling Kindersley: Gary Ombler (tl); Andy Crawford (bc).
31 Galaxy Picture Library: Robin Scagell (c, ca, r, rb).
34–35 NASA/JPL.
37 Alamy: Yendis (br). NASA/JPL-Caltech: (brr). NASA/JPL-Caltech/USGS/Cornell: (cla). NASA/JPL/Space Science Institute: (cr, crr). NASA/JPL/University of Arizona: (c).
38 SOHO (ESA and NASA): (c).
39 Alamy: John E. Marriott (br). Dorling Kindersley: Andy Crawford (b, bl). Science Photo Library: John Chumack (tr); Jerry Lodriguss (brr).
40 NASA/JPL-Caltech: (c).
41 NASA: (t, tc, tr). Galaxy Picture Library: Robin Scagell (br, brr). Science Photo Library: John Foster (bc).
42 Alamy: Matthew Catterall (tl). Galaxy Picture Library: Robin Scagell (cl, bcl). NASA/JPL/USGS: (c).
43 Galaxy Picture Library:ESO (tr).Thierry Legault (tcr, tcrb).Robin Scagell (tcl,b).
44 Galaxy Picture Library: Martin Ratcliffe (b). NSDCC/GSFC/NASA: (tl, tc, c).
45 Galaxy Picture Library: Damian Peach (bc); Robin Scagell (bl). NASA: (c, cl, cr, tr). SOHO (ESA and NASA): (br).
46 USGS: (c).
47 ESA/DLR/FU Berlin (G.Neukum): (tc). Galaxy Picture Library: Robin Scagell (bl, bc). NASA: (tr, cr). NASA/JPL/Cornell: (cra).
48 NASA/JPL/University of Arizona: (c).
49 Galaxy Picture Library: Robin Scagell (b). NASA: (tc, tr, c).
50 NASA/JPL/Space Science Institute: (c, tl).
51 Galaxy Picture Library: Robin Scagell (b). NASA/JPL/Space Science Institute: (tr, c). NASA and the Hubble Heritage Team (STScI/AURA): (crb).
52 Galaxy Picture Library: Robin Scagell (br). NASA/JPL: (cl, cr, tl, tr).
53 NASA/ESA: (bc, br). WM Keck Observatory: (bl).
54 Galaxy Picture Library: Robert McNaught (c). NASA/JPL: (bl). NASA/JPL-Caltech/UMD: (tl).
55 Dorling Kindersley: Colin Keates, Courtesy of the Natural History Museum, London (cr). NASA/JPL: (br). Science Photo Library: David McLean.
56–57 Getty Images: Hulton Archive.
58 Galaxy Picture Library: Robin Scagell (tl).
58–95 Getty Images: Hulton Archive (borders).
58–95 Till Credner (www.allthesky.com).
77 Galaxy Picture Library: Y. Hirose (clb).
79 Galaxy Picture Library: Chris Picking (tr).
96–97 Science Photo Library: David Nunuk.
98 Till Credner (www.allthesky.com): (cl).
99 Till Credner (www.allthesky.com): (cr). Galaxy Picture Library: Yoji Hirose (tl).
100 Galaxy Picture Library: Robin Scagell (c).
101 Galaxy Picture Library: Robin Scagell (cr). NOAO/AURA/NSF: NA Sharp (tl).
102 Galaxy Picture Library: NOAO/AURA/NSF/Adam Block (c).
103 Galaxy Picture Library: Yoji Hirose (tc); Robin Scagell (cr).
104 Till Credner (www.allthesky.com): (cr).
105 Till Credner (www.allthesky.com): (tl). Galaxy Picture Library: NOAO/AURA/NSF/Todd Boroson (cr).
106 Galaxy Picture Library: Damian Peach (cra).
107 Till Credner (www.allthesky.com): (cra). Galaxy Picture Library: NOAO/AURA/NSF (cl); Robin Scagell (tl).
108 Till Credner (www.allthesky.com): (tr). Galaxy Picture Library: Damian Peach (tr/insert).
109 Galaxy Picture Library: Robin Scagell (tr). NOAO/AURA/NSF: NA Sharp, Mark Hanna, REU Program (cla).
110 Galaxy Picture Library: Robin Scagell (cra).
111 Till Credner (www.allthesky.com): (tr). NOAO/AURA/NSF: (tl).
112 Galaxy Picture Library: Robin Scagell (tr).
113 Galaxy Picture Library: Yoji Hirose (tr). NOAO/AURA/NSF: (cl).
114 Till Credner (www.allthesky.com): (ca).
115 Galaxy Picture Library: Chris Livingstone (tl); Michael Stecker (cra).
116 Galaxy Picture Library: Yoji Hirose (cra).
117 Till Credner (www.allthesky.com): (cr). Galaxy Picture Library: Chris Livingstone (cla).
118 Galaxy Picture Library: Juan Carlos Casado (cra).
119 Till Credner (www.allthesky.com): (cr). Galaxy Picture Library: Robin Scagell (tl).
120 NOAO/AURA/NSF: NA Sharp (c).
121 Galaxy Picture Library: Gordon Garradd (tl); Robin Scagell (cr).
124–128 Science Photo Library: Stephen and Donna O'Meara (r).

Endpapers: NASA: ESA, M Robberto (Space Telescope Science Institute/ESA) and the Hubble Space Telescope Orion Treasury Project Team.

Jacket front and inside: Science Photo Library: Russell Croman